# MEMC … THE

# Notti…shire Coalfields



David Bell

COUNTRYSIDE BOOKS
NEWBURY BERKSHIRE

First published 2008


Reprinted 2013



COUNTRYSIDE BOOKS
3 Catherine Road
Newbury, Berkshire

To view our complete range of books, please visit us at
www.countrysidebooks.co.uk

ISBN 978 1 84674 101 2

Designed by Peter Davies, Nautilus Design

Produced through The Letterworks Ltd., Reading
Typeset by CJWT Solutions, St Helens
Printed by Berforts Information Press, Oxford

# Contents

# Acknowledgements

**I wish to thank the following individuals** for their help in writing this book: Denis Barlow, Bill Bennett, Ron Booth, Bob Bradley, Bill and Pam Buck, Bob Collier, Dr David Clark, John Deaton, Eric Eaton, Steve Hextall, Ian Gray, Mrs F. Hallsworth, Ernie Kirk, David Lee, Ken and Goff Lewis, Roy Mills, Mick Oscroft, Geoff Peace, Keith 'Scouse' Pruden, Alan Ratcliffe, Henry Richardson, Keith Staley, Keith Stanley, Henry Richardson, Colin Wyld and O. J. Young.

I am also grateful to the extremely helpful staff at the Coal Authority archives at Berry Hill, to the *Nottingham Post*, and to the *Mansfield Chad*.

# MAP SHOWING THE NOTTINGHAMSHIRE COALFIELDS

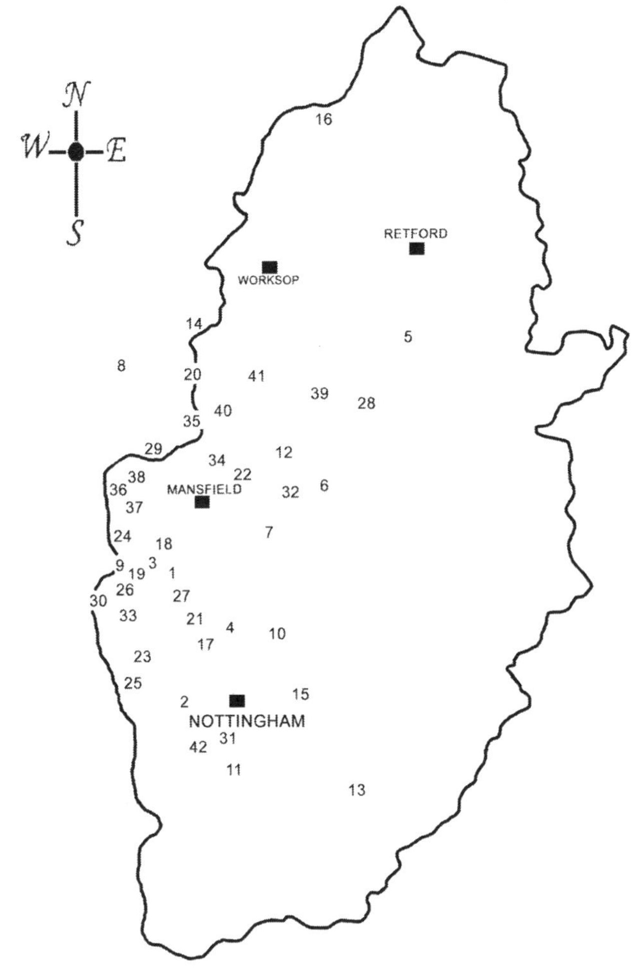


| | | | | | | | |
|---|---|---|---|---|---|---|---|
| 1 | Annesley | 12 | Clipstone | 23 | Moorgreen | 33 | Selston |
| 2 | Babbington/Cinderill | 13 | Cotgrave | 24 | New Hucknall | 34 | Sherwood |
| 3 | Bentinck | 14 | Creswell | 25 | New Langley | 35 | Shirebrook |
| 4 | Bestwood | 15 | Gelding | 26 | New Selston | 36 | Silverhill |
| 5 | Bevercotes | 16 | Harworth | 27 | Newstead | 37 | Sutton |
| 6 | Bilsthorpe | 17 | Hucknall | 28 | Ollerton | 39 | Teversal |
| 7 | Blidworth | 18 | Kirkby | 29 | Pleasley | 39 | Thoresby |
| 8 | Bolsover | 19 | Langton | 30 | Pye Hill | 40 | Warsop |
| 9 | Brookhill | 20 | Langwith | 31 | Radford | 41 | Welbeck |
| 10 | Calverton | 21 | Liby | 32 | Rufford | 42 | Wollaton |
| 11 | Clifton | 22 | Mansfield | | | | |

# Glossary of Mining Terms

**Bell Wires** Just like tram pick-up wires overhead. A pair of parallel wires under tension above the paddy, which could be reached with the ringer to stop the paddy in an emergency.

**Capple** Device fitted at the end of the winding rope affording means of coupling to the bull chains on top of the chair.

**Chair** Another term for the cage in which the miners descended the shaft.

**Cock Wood** Tinder taken home to light the fire.

**Drift** An inclined roadway into a mine that could be used instead of a vertical shaft. There could also be drifts within a mine to get from one level to another.

**Elephant's Tab**
Round-nosed shovel – likened to an elephant's ear.

**Feeder/Supply gates**
Roadways used for input of material needed for advance of the face. Also called tailgates.

**Filled off** Your nine-yard stint was all gone (in hand filling of coal).

**Gate** An underground roadway.

**Inbye** Inwards from the pit bottom to the coalface. Fresh cold air travelled inbye from the bottom of the down shaft to the coalface.

**Jotty** A small four-wheeled car which could be moved manually, usually found at the face line.

**Keps** Steel blocks which were moved into a shaft under the chair bottom to rest the chair on.

**Loader gates** Roadways used to bring coal out from a face. Also called maingates.

**Locker** A wooden or steel device inserted into the tub wheel. This locked the wheel and the friction between the wheel and the rail caused the tub to stop.

**Long Tom** A swing rail which formed a set of points to cause a tub to swap tracks.

**Onsetter** The man in charge of the cage at the pit bottom.

**Outbye** Outwards from the coalface to the pit bottom. Stale air – warm and often smelly – travelled outbye from the coalface to a second shaft – the upshaft.

| | |
|---|---|
| **Paddy** | Vehicle or net of vehicles joined together which ran on rails for transporting men to work. Thoresby had workings up to eight miles from the shaft. |
| **Panzer** | An all-metal conveyor for moving coal. |
| **Pit Bottom** | The area around the bottom of the shaft. |
| **Ram's Head** | A hand-held electric boring machine, with a handle on each side like ram's horns. |
| **Return Face** | We advance a face first, then use one of the ready-made gates to bring the whole thing back to point A. A new gate would have been previously driven to form a goal post type of effect with two gates. |
| **Ringer** | A four-foot crowbar with wedge end to lever up and pull apart. |
| **Ringer Bell** | A Y-shaped piece of gear with a hacksaw blade fixed to it, so that it would connect the bell wires together to ring a bell at the engine house. Used on haulage rope. |
| **Skip** | A long vehicle in a shaft, used for bringing coal out of the mine, like an oblong bucket with top loading and bottom unloading. |
| **Snicket** | A small excavated tunnel giving access as a walkable short cut, or a way into a face on the return. |
| **Stable Hole** | At the ends of a face, an advanced portion in front of the face line to enable the main coal cutting machine to be moved into the next cutting position. |
| **Stapple Pit** | A drift with a gradient of steeper than 1 in 1 (45%) or a shaft within the mine. |
| **Sump** | Continuation of the shaft from the cage landing to the virtual bottom of the shaft. |
| **Warwick Girder** | A girder erected at 45° between roof and floor as a stop device for runaway tubs or mine cars. |

# Nottinghamshire Pits in 1957

| *Pit* | *No. of men* | *Underground* | *Surface* |
|---|---|---|---|
| Annesley | 900 | 720 | 180 |
| Babbington | 1774 | 1447 | 327 |
| Bentinck | 1543 | 1228 | 315 |
| Bestwood | 2140 | 1776 | 364 |
| Bevercotes *(new sinking in 1957 but 10 years later there were 175 men)* | | | |
| Bilsthorpe | 1549 | 1258 | 291 |
| Blidworth | 1762 | 1452 | 310 |
| Bolsover | 1068 | 885 | 183 |
| Brookhill | 1036 | 824 | 212 |
| Calverton | 1332 | 1121 | 211 |
| Clifton | 1058 | 877 | 181 |
| Clipstone | 1428 | 1150 | 278 |
| Coppice No 1 | 700 | 576 | 124 |
| Coppice No 3 | 780 | 642 | 138 |
| Cotgrave *(developing in 1957 with 69 men but 10 years later–1116 men)* | | | |
| Cresswell | 1449 | 1140 | 309 |
| Gedling | 2372 | 1911 | 461 |
| Harworth | 1934 | 1546 | 388 |
| Hucknall | 1523 | 1160 | 363 |
| Kirkby | 1618 | 1314 | 304 |
| Langton | 965 | 744 | 221 |
| Linby | 1503 | 1139 | 364 |
| Lodge | 445 | 362 | 83 |
| Mansfield | 1704 | 1434 | 270 |
| Mapperley | 778 | 617 | 161 |
| Moorgreen | 1511 | 1211 | 300 |
| New Hucknall | 637 | 490 | 147 |
| Newstead | 1251 | 982 | 269 |
| Ollerton | 1529 | 1246 | 283 |
| Pleasley | 1331 | 1060 | 271 |
| Pye Hill | 1024 | 816 | 208 |
| Radford | 598 | 487 | 111 |
| Rufford | 2108 | 1743 | 365 |
| Sherwood | 1257 | 986 | 271 |
| Selstone/Silkstone | 683 | 568 | 115 |
| Silverhill | 1263 | 1016 | 247 |
| Sutton | 605 | 467 | 138 |
| Teversal | 912 | 761 | 151 |
| Thoresby | 1780 | 1457 | 323 |
| Warsop | 1585 | 1236 | 349 |
| Welbeck | 1600 | 1282 | 318 |
| Wollaton | 711 | 602 | 109 |

# Introduction

**All the memories in this book** were related to me by miners who had worked in the Nottinghamshire pits, from Creswell to Gedling, from Clifton to Ollerton, from Newstead to Sherwood. Many of the reminiscences are humorous – the miners' banter and jokes, and the tricks played between comrades down the pit. Other recollections were more serious, stories of accidents and the conditions in which the men worked. Together, these memories paint a picture of the life of the tough but resourceful Nottinghamshire miner.

The Nottinghamshire coalfield is part of a vast coalfield that stretches west into Derbyshire and north into Yorkshire. In general, the western part of the coalfield is closer to the surface than further east. Not surprisingly, the first collieries were situated in the west towards the Derbyshire border, where the coal seams were nearer to the surface. Further east, Harworth and Bevercotes pits both had shafts that were more than 1,000 yards deep, among the deepest in the country.

Although coal had been extracted in Nottinghamshire for many centuries, many of the larger collieries in the county were sunk in the 20th century. These include many sunk in the 1920s and 30s, as well as Bevercotes, founded in 1954, and Cotgrave in 1964. This meant that Nottinghamshire was one of the leading coalfields in terms of new technology. Thoresby colliery was the earliest mine in the country to be completely powered by electricity, and it was the pioneer colliery in terms of mechanised production. It was also the first pit to turn over a million tons of coal per year and, by the late 1980s, it was producing over two million tons of coal.

Because of the high production and modern technology employed, miners came to Nottinghamshire from older coalfields

*Drilling in the shaft bottom at Cotgrave colliery. (The National Archives, Crown Copyright)*

in other areas. Jock Purdon, a miner, a poet and a singer from Chester-le-Street in County Durham wrote: 'Leave your picks behind ye, ye'll no need them agen. And off you go to Nottingham, join Robin's merry men. Leave your cares behind ye, your future has been planned. And off ye go tae Nottingham, tae Robin's Promised Land'.

As well as miners from other parts of the UK, Nottinghamshire attracted miners from further afield: Polish, Latvian and Lithuanian miners were just some of those mentioned to me by several Nottinghamshire miners who worked alongside them. Colin Wyld told me, 'Just after World War Two, there were lots of displaced persons – Poles, Latvians, Rumanians – living in Edwinstowe, and Thoresby colliery employed quite a few of them. Herbert Meredith was the deputy whose job was to enter their names and check numbers into his time book, for them to be paid. Their names were so complicated, with Z following P or N, that Herbert could not spell them. He used to write in his time book a list, which looked like this: "A Pole–1462, A Pole–1298, Lat–1470, A Pole–1471" and so on.'

There were several collieries situated on the Nottinghamshire–Derbyshire border, and many men crossed the county boundary to work. Langwith pit was in Derbyshire but it was so close to Nottinghamshire that the county boundary ran down the road between the terraces of miners' houses. Two other pits that were actually in Derbyshire were classed as Nottinghamshire collieries. One was Creswell and the other, very surprisingly, was Bolsover which is situated five miles into Derbyshire. The reason why these are Nottinghamshire pits dates back to the miners' strike of 1926. In that year, the miners were on strike to prevent their hours being extended and their wages cut. For a brief period, the TUC supported the strike by calling out transport workers, engineers and others in a general strike. However, after nine days, the TUC caved in to government pressure – the general strike was being called a Russian-style revolution – and the miners were left to struggle on alone for nine months. Following this, a Nottinghamshire MP, George Spencer, set up an alternative union for miners called the Nottingham and District Miners' Industrial Union. This union,

*Men coming off the cage at Bevercotes colliery. (The National Archives, Crown Copyright)*

*Underground junction of roadways at Manton colliery. (The National Archives, Crown Copyright)*

usually referred to as the Spencer Union, vowed to work closely with the mine owners, never to strike, and to stay out of politics. It was often referred to by miners as 'the scabs union'. Although many miners hated the Spencer Union, the mine owners loved it, and many mine owners from neighbouring non-Nottinghamshire areas declared their mines to be Nottinghamshire pits and thus it was that Bolsover and Creswell became Nottinghamshire collieries.

History repeated itself in 1984-85 when, during the miners' strike to prevent collieries being closed down by the government, the majority of Nottinghamshire miners chose to work. Some 7,000 Nottinghamshire miners came out on strike while 27,000 worked and this split is still felt today, even among ex-miners. You are either one who supported the strike or one who worked. Just as in 1926, the split led to the formation of a breakaway, less militant union – the Union of Democratic Miners (UDM).

Many of those who worked or 'scabbed' and who joined the UDM hoped that the government would reward them by keeping all the Nottinghamshire pits open. However, they were disappointed and twenty-five pits were closed. Roy Lynk, a founder member of the UDM and one of its leaders, said that the government had stabbed him in the back. The members of the NUM – and Nottinghamshire still has a large number of members who retained their NUM membership – may be forgiven for thinking, 'We told you so'.

Nottingham, however, is one of the very few coalfields to still have two pits working, though they are now under private ownership. These are Thoresby and Welbeck collieries, now owned by UK Coal. A third Nottinghamshire pit, Harworth, was closed or 'mothballed' in 2007, but UK Coal have announced that, because of the soaring price of coal, it may well be reopened. Welbeck colliery is due to close at the end of 2009, and the miners there could transfer to Harworth if it is re-opened. After a deal with the power firm EDF, UK Coal have said that Thoresby colliery could be turning coal for at least another ten years.

Keith 'Scouse' Pruden – always known as Scouse because of his Liverpool origins – worked in various Nottinghamshire pits, including Rufford and Annesley-Bentinck, as well as at pits in Yorkshire and Derbyshire. He confirmed the current resurgence of

*Headstocks at Welbeck colliery in 1965. (The National Archives, Crown Copyright)*

*The boiler house at Annesley colliery, 1973. (The National Archives, Crown Copyright)*

interest in coal mining, telling me, 'Somebody did contact me about going back down the pit, just a year ago. I got a phone call. "Is that Keith Pruden? My name's so-and-so, and I work for UK Coal. I just wondered – are you working at present?" I said that I was. "Oh right. Have you ever considered going back down the pit?" I said, "I just have. I've just considered it and it took me about a second." However, he went on to explain that UK Coal was recruiting for coalface workers and coalface supervisors. He said, "Apparently you can train people on the coalface," – which I can do – "and I wondered if you would be interested in it." I said that I wasn't interested. I didn't find out which pit he wanted me to go to, but Welbeck colliery is recruiting at the moment. They've just set on 55 coalface and development workers.'

Keith then added, 'But also – and I know this is an absolute fact – they've done test boring at Teversal last year. Teversal wasn't a very deep mine, and apparently there's an outcrop that they can actually use to create a drift mine. Deep shaft mines are a bugger to maintain, but drift mines are cheaper to maintain than shafts. So now they're looking at collieries where they can do drift mines. The thing is with a drift mine, you'll drift in with the natural gradient of the coal, so while you're developing, you're turning coal. So it's paying for itself. This is the way they're going now. Do you know that there's thirty billion tons of coal still not been mined in this country?'

So, it may be that coalmining is set to continue in Nottinghamshire for some time to come.

# Chapter 1

# Starting Work

**The miners who talked to me all described** what it was like to start work at a colliery, to go from the world of childhood into another world. Ron Booth explained, 'I grew up in the Radford area of Nottingham. At Easter 1947, I left school at the age of fourteen to go to work. I wanted to go into an apprenticeship, but my elder brother was already an apprentice and I was told that my parents could not afford another one in the family. My dad found me a job at Sketchley Dye Works, and I was there about three months. My dad had been a coal miner at one time but in 1937 he got buried and decided to leave the pit. He didn't want me to go down, but I went and signed on. It was 23rd June 1947 – my sister's birthday. Along with three schoolmates, I went to Babbington colliery. We all passed the medical and reported to the pit top on the Monday morning. After a few weeks working in the coal screening plant, we were sent to Bestwood Training Centre.

'There we had everyday schooling and, in the afternoons, we were taught various mining practices in a mock-up coalface under the building. Every Wednesday, we all went back to our various collieries to be shown round the different pit top workshops where the tubs, electric cables and machinery were repaired. Then we went down the pit to see the various places of the workings of

*Demolition of the headgear and winder house at Babbington colliery, 1956. (The National Archives, Crown Copyright)*

*The mono-rail riding system at Babbington colliery. (The National Archives, Crown Copyright)*

*No. 4 shaft at Babbington colliery, pre-1958.*
*(The National Archives, Crown Copyright)*

the mine. We met the training officer at dawn on a cold November morning outside the lamp cabin where we were supplied with an electric bucket lamp, weighing about nine pounds. We were kitted up with helmet, knee pads and the most important piece of equipment, a water bottle. We went to the gantry, passing the hissing and clanking of the monster steam engine which worked the cages, bringing the coal to the surface in tubs. These cages held anything up to one ton in weight.

'In the distance, screeching overhead buckets were tipping the slag on a pyramid of waste. We got to the shaft side where the banks man searched us for combustible materials. We were then crowded into a packed narrow cage like herrings in a barrel. It smelled of sweat, dampness and danger. The gate fell to seal us in. The cage just moved upwards slightly to clear the safety jacks and, without warning, we were dropped like a stone into the bowels of the earth. For a moment, I thought I had lost my stomach. Had the cable snapped? My companions continued to mutter to each

*Miners in the cage going on shift at Calverton colliery. (The National Archives, Crown Copyright)*

other, while the glistening cables, slimy beams and flashing lights passed by. At about 1,000 feet, there was an awful jolt, which gave me the impression that the cage was rising again. A dusty burning smell filled the air.

'We stepped out of the cage into the pit bottom. The grime-covered night shift stood waiting to take our places in the cage. A strong draught of air came from the rattling ventilating machines and the noise of the tubs bumping into one another, as they were lined up behind the squeezers, ready to be released and to be pushed onto the cage when coal turning started up again.

'We made our way up the main road following the empty tubs, heading for the loading point, stopping at brief intervals along the way, as the road narrowed when the tubs full of coal passed the empties going in the opposite direction. After about two miles, we entered the return air road of the coalface; this is the road where pit props and other materials were supplied to keep the coalface working. We went through two big wooden doors to be greeted by warm sickly air. All the air that had come through the coalface was now coming down this road, making its way out of the mine. The air was obnoxious, smelling of explosive smoke mixed with horse manure.

'At this time, I thought my arms must have stretched two feet, with the weight of the bucket lamp. I kept changing it from hand to hand; but it was quite a struggle to walk with it. We went up the road in single file, led by the training officer. I had to turn and duck as the steel girders holding the roof up were bent and twisted. Ominous cracking noises came above my head and I wondered how much further there was to go, as my back, neck and thighs were aching. I wasn't used to heavy steel toe capped pit boots; they seemed as heavy as divers' boots. I was trying to see where I was walking on the uneven ground, whilst watching the low girders at the same time and trying not to bang my head. I had the first feeling of claustrophobia, and the glamour of working down a coal mine was beginning to fade.

'As we walked up the road, we were met by the ganger lad coming the other way, sitting in the empty tub and driving the pony. The training officer remonstrated with him, telling him to get out and walk, and to lead his pony. I don't think he took much notice of him because, as we walked on, I turned round to see him get back into the tub to continue his ride. We eventually reached the coalface and the rippers were digging into a great pile of waste, which they were packing in each side of the road. These men were responsible for extending the roadway when the coalface moved forward each day after the extraction of the coal. Through the dust and noise I watched a shadowy group of men; some wore only shorts, boots and knee pads, sweating as they went about their task. It must have been 80° F. After about thirty minutes we were told it was safe to enter the coalface and we began to crawl on our hands and knees for 200 yards along the face, listening to the banter from the face workers as we passed them.

'When we arrived in the Main Gate, we were completely exhausted and thirsty, and we just collapsed as we sat down. Some time during the shift, the pandemonium stopped, and some of the face men came to eat their snap. We were told to eat our snap – sandwiches – and wash them down with a little water. We were told not to gulp it, only sip it, or we would end up with stomach cramp. I have never been so thirsty in my life as I was that day.

'We then made our way back to the pit bottom, going through endless roads and turns. There was no conversation as we watched men pushing and tugging the full coal tubs onto the cage. No sooner was that done, in a bat of an eye, the cage was on its way to the surface. They could wind over 200 winds an hour, which was some feat, hard and strenuous work. These

*Shotfirer drills a shot hole at Calverton colliery. (The National Archives, Crown Copyright)*

men were not beefy, but wiry men, tough and strong as horses.

'When we were allowed to enter the cage to return to the pit surface, we were all deep in thought about the experience we had just had. Light was coming towards us as we reached the top. There was a clank as the cage passed through the safety jacks and settled on them, and finally we shuffled out into the sweet air. When I arrived home, my mother had got the tin bath full of warm water ready for me to bathe. I was tired but I was also very proud to have become one of the unique and fearless men they called miners.'

Bob Bradley had experienced the traumatic thrill of going down a mine when he was a young boy: 'The first introduction to mining was on a Sunday morning when a friend of my Uncle Jack's who was an ostler at Moorgreen colliery asked if I could go with him to feed the ponies. We duly went along, my uncle and this chap, both wearing flat caps. I'd got my blue gabardine coat on. I was 10 years old, so they found me a round helmet, very soft, one that would have been worn by an overman or an undermanager. Anyway, the engine driver was tipped the wink that there were two newcomers on board the cage, so we shot down the shaft at what seemed to be the speed of light. One gets a strange feeling when you get halfway down the shaft; you feel as if you're coming back up again. And, of course, you lose your hearing to some degree. Your ears pop, and sometimes you can become slightly deaf. You need to gulp, pinch your nose, or chew something. And the other thing is that you can get a feeling of nausea or dizziness the first time you go down. And the deeper it is, the more times you need to gulp or pinch your nose. Anyway, they took me round the working. We ducked and weaved all the way to this coalface. I remember that every little piece of coal seemed to dig into my knees, because we'd not got kneepads on, and I seemed to crack my head on every low girder there was. We made our way out to the pit bottom, and I enjoyed feeding the ponies with the odd apple and carrot we'd taken down. I came out of there and I said, "You'll never get me to set foot on a chair again."

'Six years later, in early 1953, I started work down the pit at Silverhill Training School. One had to be there for 6.30 in the morning. There were no bathing facilities for us, because there were no spare lockers. Silverhill pithead baths were shared by miners from Teversal, because originally they were owned by the same company, Stanton Iron Works. I was doing sixteen weeks underground training which new entrants to the mine, at 15 years old, had to do. In fact, I was sixteen, having left grammar school, but of course I

*A miner underground at Moorgreen colliery. (The National Archives, Crown Copyright)*

*A miner crawling under 2 ft 10 ins low seam chocks at Moorgreen colliery. (The National Archives, Crown Copyright)*

had to go through the same training. Boys of fifteen who had finished their training were earning about £4 a week, but I was set on at £2 15s a week. There was no overtime pay, even for working Saturday afternoons or Sundays. My hours of work were defined as: "You will attend work at whatever hours are laid down from time to time by the area surveyor."

'I was doing the underground basic mining. You learnt how to push the conveyor over on a static coalface – a small coalface about 50 yards long. You did a bit of shovelling, and building packs. The following week you took it all to bits again. One team took it off and then you swapped over. There was a little bit of coal to work. It was supposed to be static but you took a yard of coal off every now and again. We all shovelled like mad to try and get as dirty as possible, to look like real miners. After this, you had two days at college, and three days down the pit learning all the

*Looking inbye along a trunk conveyor road at Clifton colliery. (The National Archives, Crown Copyright)*

*The screening plant at Gedling colliery. (The National Archives, Crown Copyright)*

rudiments, like how to fit a pony up with its limmers, also all the trappings of underground haulage work.'

Bob had a bit of a mishap on his last day of training, which caused several miners to get covered in a most unpleasant substance. 'They gave me the task of bringing out the latrine tubs, bringing them out to the pit bottom. The latrines were at the top end of this training gallery. They emptied the latrines at the end of every sixteen weeks, so you can imagine what they were like. We put as much dust as possible into these tubs. The other lads were given the task of putting all the latrine material into these tubs, and I got the job of lowering them down because I was the best at clipping on. I put the star clip on at the back, tightened it up, put a Smallman clip on at the front. They started the engine up and off we went down this one in forty grade. I took the front clip off, put the lockers in, wooden lockers, in the front wheels as well as the back wheels, and took the star clip off. And the tubs started to slide quite heavily. I put more lockers in and, as they went round a bend, the tubs started to gather speed. I was quite deft at putting these lockers in, but as I flicked them in, each one snapped like a carrot. They were oak pegs but they were so old and dry, sort of wood-wormy, that they snapped. These tubs started to hurtle down towards the pit bottom, which was about 50 yards away. I started waving my light and shouting. Fortunately the men could see what was happening, and they got out of the way. Fortuitously, you might say, there were about a dozen tubs resting on the jack-catches, as we used to call them, and our latrine tubs ran into them. The back one flipped up in the air and all the detritus was flirted up and covered everything. I went to ground, I did. The smell was there for days. That was my last trip down there, but I was completely exonerated, because it was found that I hadn't missed putting the lockers in any wheels. These wooden lockers were not up to the job, so it was decided that in future they would only use steel lockers.'

Lads who lived in Nottingham would have more choice of workplace when they left school, including the famous Raleigh cycle factory, and Players where cigarettes were manufactured. Bill Buck told me, 'I worked in the Raleigh industry when I left school, and I went down the pit in 1962, when I was about eighteen. That was at Clifton colliery, but first I did nine weeks training at Hucknall. The first time I went down the shaft it were frightening. Your ears go deaf through the air pressure. My first job was on the stone dust barriers, near the coalface. They support the rings and you have to keep building the stone dust on them, keep chocking up. When you're doing the stone dust barriers, the dust is like fog, and you're breathing all that in.

'Then I worked at the pit bottom, on the empty tubs. You get one shaft with the clean air going down, and another with the warm air coming up. I was working where the cold air came down. It was cold down there, freezing. When the cage came down with the empties on, I had to make sure they stayed on the track, pushing them round a bit.'

Keith 'Scouse' Pruden had wanted to go into the Royal Navy,

*Pit bottom showing 15 cwt capacity tubs at Linby colliery. (The National Archives, Crown Copyright)*

*Cable belt roadway with arch girders at Linby colliery. (The National Archives, Crown Copyright)*

but found his maths wasn't good enough so he decided to go down the pit. 'My second choice was to go in the pit, because my step-dad worked there as an electrician. This was up in Wigan, but then we moved down to Nottinghamshire because up in the north-west, the pits were coming to an end and Nottinghamshire coalfield was expanding. They were setting on more and more, plus the housing that they'd built down here was better. So my step-dad transferred to Bilsthorpe colliery, and we came here as a family. I'd be fifteen and a half.

'I wanted to train as an electrician but after the exam they said, "Well you've not got enough to be an electrician but you've got enough to be a manager." So they took me on as a management apprentice, which is a mining craft apprenticeship, with day release to college. Then my two brothers – one a year older than me, the other a year below me – they went for their exams, and all three of us ended up with mining craft apprenticeships. We all joined the pit within two months of each other, and we all worked at the same colliery, Rufford. It was a lowish rate of pay, £4 0s 6d

*Rufford colliery in 1987. (Keith Pruden)*

per shift. You couldn't do overtime, and you couldn't do weekend work, not until you were eighteen or over.'

The usual process with lads joining the pit was for them to work on the surface until they were sixteen, then work underground in the area at the bottom of the shaft, the pit bottom. Scouse followed this procedure. 'I started off on the surface repairing coalface props. I had an instructor called Frank Ulliot, who had been a deputy underground, and he was my first instructor. There was no undercover repair area, it was all done out in the open air. So whenever it was chucking it down with rain, I used to have to go out and do all the repairs. I used to get wet through. If it was a lovely sunny day, Frank would come out and help me. But he'd never come out in the rain.'

Ernie Kirk's start was the same, on the pit bank, the surface. 'I went to work at Gedling colliery straight from school. I worked in the stores to start with, and then I moved to the screens, still on the surface. I went next from the surface down to the pit bottom, working on the roads, then eventually I went onto coalface training. In the meantime, I took my shot-firing ticket, but they didn't send me from the Coal Board, I went on my own bat. I used to go to night school at Arnold. I finally got my shot-firing ticket. I was on the coalface, no mechanisation then, all hand-filling. Your task was nine yards of coal in length. They used to say nine yards, three foot six high, five foot undercut, made sixteen ton of coal. So I was on that for a few years, and then I went shot firing in 1952. I was shot firing for a year or two, and then I went deputying, and then I went as an overman. The last ten years I was senior overman at Gedling. My whole forty year was at Gedling colliery.

'The first time I ever went down was while I was working at the stores. The storekeeper was going down to the stables – because it was all pit ponies then – and he said, "Would you like to go down?" I said, "Yes, I would." Of course, my father had worked down there more or less all his life. So we went down just to visit the stables in the pit bottom, and check on the stock, the food. The ostler was there that used to look after them all, and of course

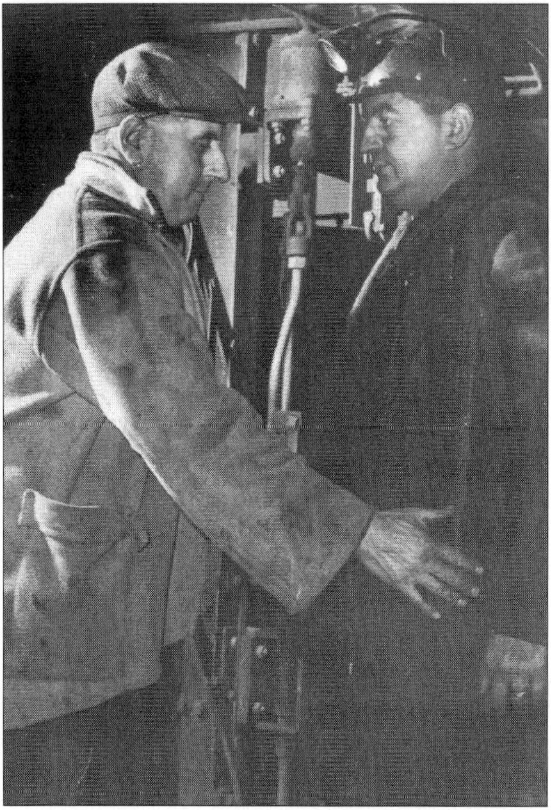

*Frisking a miner for cigarettes and matches before he goes down Calverton colliery. (The National Archives, Crown Copyright)*

the lads that were working. They used to pick up their ponies in the morning, and take them inbye, material handling, then drop them back in the stables at the end of the shift. That was the first time I ever went down, when I worked at the stores. The shaft at Gedling was about 400 yards deep.'

Colin Wyld's recollection of his early training at Lound Hall, near Bevercotes pit, mentioned the low starting wage. 'I remember the first ever wage packet I drew at Bevercotes pay office was for £4 19s 11d. When I asked why did they not put in one more penny to make it a round fiver, the man told me to sod off. So I did.

'It was early days at this new training complex and it wasn't finished yet, so we spent quite a lot of our time there building imitation underground roadways, which were of course above ground, with dim or no lights, so that we had to get used to working with a battery-powered cap lamp. These training galleries, as they were called, were authentically put together, with haulage ropes, engines, rails and tubs, etc. What an adventure it was for the younger ones like me. It was grotto land. The training was to go on for sixteen weeks, by which time we were all ready and anxious to get on to the mother pit, for which we had been trained. What a mistake that was! We didn't realise where we were best off. Looking back, it was a holiday, though we were not afforded bathing facilities. We had to use public transport, looking like something from the *Black and White Minstrel Show*, and wearing hobnailed boots which made a terrible clatter as we went to the top deck for a crafty fag, purchased singly from the pit canteen for 2d each.

'Training was now over, and we were going to be thrust into the harsh reality of actually earning a living on our own merit. We were going for the big time in adult life. But we were novices, beginners, and they used to say we didn't know hay from a bull's foot.' Despite not knowing – in his wonderful phrase – hay from a bull's foot, Colin was to spend the next thirty six years at Thoresby Colliery.

## Chapter 2

# Banter and Backchat

**Groups of men who work together** often have a sense of humour all their own. In the pit, with its often difficult conditions – men may be working in the cold of the pit bottom or in the heat and humidity of the coalface, and may be working in a seam only 3 ft high – a sense of humour helps to keep them sane.

Some of the humour is verbal banter, backchat between the men. Keith Stanley, the national vice-chairman of the NUM, as well as the Nottingham NUM general secretary, recalls one such occasion from his time at Newstead colliery, when he and his mates were trying to get some new equipment to the coalface. 'One of our laughs – well it weren't a laugh for me – was when we were getting this gear head in. The gates were that low we couldn't get stuff to the coalface, so we had to handball it, had to carry it. We'd got to get it up this dint and up to the face. We were doing it with a five-inch ramp and an old haulage chain dogged onto it, and pulling it up under pressure. The gear head got fast and I was in the middle between the gear head and the ramp, when the chain broke. It flew and shot round my foot,

*Cutter and jib about to start a cut at Calverton colliery.*
*(The National Archives, Crown Copyright)*

chained it. It actually broke four or five bones on the top of my foot. I dropped to the floor in agony, hands covering my face and the lads came running up to me. They said, "It's hit him in the face," and pulled my hands apart. Three of the lads screamed and one said, "Oh my God, what a mess it's made!" I said, "It's my foot, you twat." That's the humour of the pit – sarcastic humour.'

David Lee told me a story about an old miner whose nickname was Tush. David recalled, 'Back in the old days, Wigfalls the electrical shop, used to sell items on HP, and a chap named Arthur Slater was their rep for Langwith village. On Mondays and Tuesdays he used to come round house to house in the village to collect the three shillings or four shillings for the next instalment. He'd knock on a door, and it'd often appear that there was nobody in, apart from the fact that the curtains were twitching until he'd moved on. When Tush retired, somebody asked him how he was getting on with nothing to do. Tush said, "I'm not fully retired – I've got a part-time job." "Have you, Billy?" they said. "So what's this part-time job?" "Well," he says, "on Mondays and Tuesdays I'm going to follow Arthur Slater round and sound the all clear when he's gone!"'

Backchat also came in useful in putting one over on the manager or the deputy, bringing the authority figure down to size. Geoff Peace, who worked at Blidworth pit, told me, 'I remember when one of the old lads was having a nap. It happens when you get up early in the morning, you know, you go down the pit dead tired. This old lad, he was bloody tired and his head went down. Everybody seen it, but then the manager comes up the gate and he kicked him. He woke up and the manager said, "Don't you know who I am?" The old lad says, "We're bloody all right here. We've got people wandering down the pit and they don't know who they are!".'

Sometimes the banter could contain an element of wisdom. Geoff continued, 'There was these colliers sat on the paddy one night, they were discussing how far the naked eye could see. One was saying, "Well, I think I've heard somewhere that when you stand on the beach and look out to sea you can see about fourteen

*Steam winder at Blidworth colliery. (The National Archives, Crown Copyright)*

*The canteen at Sherwood colliery. (The National Archives, Crown Copyright)*

miles to the horizon." Suddenly this old bloke come up and gets in the same paddy car with them. They said, "George, how far can you see with your naked eye?" He says, "When it's dark, I can see the moon!" Very true when you think about it.'

Sometimes the banter could be heard in the canteen. This story is again one recalled by Geoff. 'When you'd been on night shift and you stopped two hours overtime, you got what they called a meal ticket. If you took that meal ticket into the canteen you could have a meat pie, or a couple of bacon cobs or sausage cobs. Now this lad called Dick, he goes in and he says, "I'll have one of them pies." Now they had a canteen manageress that was so severe she ought to have had a bolt through her neck, so mean that she had a teabag count every morning. She was that bad, they used to call our canteen the ulcer factory. Anyway Dick gets this bloody pie and he opens the pie up and he's looking at it. She says to him, "What you looking at, Dick?" He says, "I'm looking at this. There's a bloody great big grub in here." She looks and says, "That's not a grub, Dick, that's fat." He says, "It ought to be bloody fat, it's ate all the meat." That story, and it's perfectly true, became a standing joke at Blidworth pit.'

Another tale where the butt of the joke was an authority figure was told to me by Mick Oscroft who worked at Shirebrook colliery. 'One time, a pit undermanager, who was not liked by many because he was bad-tempered, arrogant, spoke if he wanted to, ignored you if he didn't – you know the type – was looking for the district deputy, a chap called Jack Spalding, and he couldn't find him. The undermanager's temper was building by the minute, when he stopped at a tandem point to enquire as to whether the attendant knew of his whereabouts. "Do you know where Spalding is?" he asked. The attendant not being too bright replied, "Aye, I think it's in Lincolnshire, in't it?" I don't have to tell you the response from the undermanager, but they say they could hear him shouting from as far as the pit bottom.'

Bob Bradley, a mining surveyor, remembers a couple of incidents from Ollerton colliery, where wit and humour was used to "break" an undermanager. He recalled, 'There was a character called Lijah, whose proper name was Nigel. He used to do dintings with his friend Peter, bits of back lips and things like that. One day this new undermanager went down the north side, and we'd got some idea what yardage they'd been doing. Lijah used to have a board on the end of the gate, and there were two sticks chalked on it. The undermanager says, "You're doing very well. Eleven tubs? Very good." He noted it down and off he went. He'd gone about half a mile when he suddenly thought "I didn't see that many tubs". So he went back, and says, "Where's these eleven tubs?" Lijah says, "Here." There's just two tubs. The undermanager said, "You said you'd got eleven." Lijah goes, "One and another one next to it, when I was at school, that made eleven." So he'd broke him. The undermanager said, "Right you two, we're having you on three shifts." He meant days, afternoons and nights. Lijah says to him, "Thank you very much. I have prayed for that day. Since I started work I've always wanted three shifts." The undermanager says, "You like the idea of three shifts?" Lijah says, "I certainly do." "All right, " he says, "you start dinting on number 2s airway on the Monday night shift." Lijah says, "I'm not doing no night shift, what are you talking about?" The Undermanager says, "Well, you've just agreed to go on three shifts." Lijah says, "I have: – I'll do Tuesdays, Wednesdays and Thursdays, days regular." The undermanager was broke again.'

Henry Richardson, the legendary general secretary of Nottinghamshire NUM, told me a true story which shows how some miners did not treat their wives with much sympathy and consideration. 'At one time, when your free coal was delivered, it used to be tipped outside so you had to take it in yourself. Then things changed, the Clean Health Act and that sort of thing, and coal began to be delivered in bags. This had been going for four or five years, but then we had this real severe winter. And they got behind with the coal deliveries, really miles behind. So we got a special dispensation from the council to start delivering lorries of loose coal to try to catch up. This meant that once again the coal had to be barrowed round from the road to the back of the house. And we had this bloke come in to the union office steaming and complaining. "Bloody hell," he said. "We can't have this. It's killing our missus's back." And he was serious! It caused a big laugh. Breaking-our-missus's-back, we used to call him.'

Colin Wyld, who worked at Thoresby Colliery for 38 years, described how one man set up a little business at work when he was not busy. Colin explained, 'Thoresby was billed as the world's first all electric pit, with electric winding engines, powered originally by boilers driving steam turbines, which in turn drove the motor generator sets, making direct current for the winding engines. No. 1 winder was the main riding winder and as such had a fair bit of downtime – un-busy time – so one of the engine men, Ralph Richardson, occupied his time by giving haircuts to the people that were in the know. The winding cab was only about ten feet square, very cosy and very private. One day, however, the manager, Mr Thorneycroft, happened to come into the engine house and caught Ralph in the act. He said, "I pay you to wind, not to cut hair. Explain yourself." Ralph retorted, "Sir, his hair

*Underground electrical maintenance at Bevercotes colliery, August 1967. (The National Archives, Crown Copyright)*

had grown in company time so why can't I cut it in company time?" The manager replied, "It didn't all grow in company time, did it?" and Ralph said "I know, so I'm not cutting it all off." He was, of course, reprimanded and warned that this must not happen again. Ralph made sure that it wouldn't by getting a bolt fitted to the inside of the sliding door of the control cabin!.'

Mick Oscroft's final tale demonstrates the quick devastating wit of the Nottinghamshire miner. Mick, still an apprentice, and a fitter were making their way inbye and had stopped for a minute with the tandem man. 'As the fitter and I chatted, the tandem man busied himself cleaning up some spillage. The fitter was telling me of an altercation in Shirebrook market place at the weekend. Apparently a local man, well known for his temper, especially in drink, and also for his ability not to have had a job since leaving school, had been fighting and making a nuisance of himself. Towards the end of this conversation, the fitter said to me, "It took six policemen to put him in a straight jacket." And as quick as a flash, without looking up or breaking the rhythm of his shovelling, the tandem man said "Ah! It would have took twenty to put him in a pit donkey jacket!".'

I will let Geoff Peace have the final tale of this section on banter and backchat. He recalled, 'In the morning, when the deputy manager used to come to the pit about seven o'clock in the morning, he used to get to know all what was going off, so he could report to the manager when he came. Well, this particular morning, Jack Bould, the deputy manager, gets on to the telephone exchange and there was a lad called Albert Thompson. This is a genuine story I'm telling you – he gets on the phone and he says, "Albert, how's 1-16s doing?" He says, "Well, Mr Bould, they haven't been onto me yet. They haven't told me anything." Jack Bould says, "50s, where are they? What's the position on 50s?" Albert says, "Well, they've got a bad joint on the loader. I do believe they're going to run it round and they're going to change it." "What's 8s doing, Albert?" "No there's nothing, Mr Bould. They've set off from the loader gate, and I haven't heard of any progress. I don't know what they're doing." Just at that minute the colliery cat walked by the deputy manager's office, and he says, "Albert, I've got a cat here that knows more about the pit than you do." So Albert says, "Right-oh, Mr Bould."'

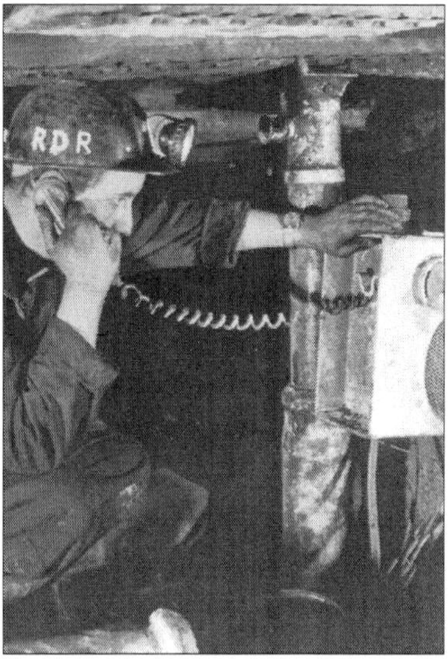


*Miner using the coalface telephone at Bevercotes colliery. (The National Archives, Crown Copyright)*

'Three weeks later, with the shift cycle, Albert Thompson's back on days again, and Jack's doing his normal round. "Albert," he says, "how's 1-16s doing?" "Can't tell you owt because they ain't been on to me yet." "And 50s?" "50s is set off from the loader gate and they're doing all right." "And 8s?" He says, "Well they've got a bit of a breakdown in the middle of the bank. They're going to let me know when they start up." And Albert says to the deputy manager, "I realise, Mr Bould, that this information ain't quite what you wanted, is it? But I've got a cat here. Would you like a word with him?"' I think the moral of the story might be that management really shouldn't try to out-banter the men.

# Chapter 3

# Nicknames and Odes

**One feature of the pit was the nicknames** that miners gave to each other. This was quite useful when there were several members of the same family working together. David Lee remembered, 'There was a Parker family in Langwith, about four or five brothers and they all had pit nicknames and we used to call them Hug, Shug, Fug, Nunk and Nander. I think they're all dead now, and the last one, George, who was Nunk, died when he was 93. And when they said Hug, Shug, Fug, Nunk or Nander Parker, you knew exactly which one they were talking about. Whereas if they'd have said George or Bill, they'd say, "Bill? Which one's he?" I'd say, "Shug!" and they'd go "Oh aye, Shug."'

Keith Pruden contributed, 'Everybody had nicknames. There was Cudder and Budder. I can remember Flea – he used to buzz round like a bloody wasp – and Chippy. Chippy was a good bloke. He was older than me, about 60 year old. There was Gonzo, named because he looked like that thing off the *Muppets*. I remember Goffy, he was one of the old lads. He was in the pit

*Electricians joining a cable underground at Clipstone colliery. (The National Archives, Crown Copyright)*

when it was ball and fire, everything on the shovel, when they used to have to undercut the coal.'

Ron Booth told me, 'Nicknames? There used to be one man called Chucker Harris, and that was because of the way he worked. He used to throw the coal into your pile. Another man was nicknamed Barren, and that was because he had six or seven kids. And one was called Chatterway, because he'd never stop talking. Blah, blah, blah. I expect Chris Chattaway, the runner, was around at the time. But you'd sit there waiting for the manrider and you could hear him yapping everywhere, he never stopped.' Keith Staley had a similar nickname. 'On nicknames, they called me Zatopeck – Emile Zatopeck was the runner who got about seven Olympic gold medals in the 1950s – because everywhere I went I used to run. I used to run to the pit bottom, and we used to play shove-halfpenny in the pit bottom until it was two o'clock, to get wound up out of the pit.'

Colin Wyld remembered vividly the characters – some rather unscrupulous – that he met while training at Crown Farm. 'We did have to undergo actual conditions underground, so after a while we were transferred to a colliery called Crown Farm, which was really Mansfield colliery. Anyway this is where we met the hard core of training instructors, some of whom were kindly and some who were not and tended to take advantage of the new recruits. One such character was called Hackman, which may or not have been his real name. He used to con us out of our small change or sweets or whatever he could get. We soon got to know who was good and who was bad, and the ones we disliked were given nicknames which ranged alphabetically between Adolph and Zombie. Names in between included Sandwich Swiper, Spangle Bandit, Bait-box Raider, and so on. As you may have gathered, some of these guys were pretty obnoxious and used to steal from our haversacks, which had to be left at a designated place.'

Bob Collier told me, 'There was one chap, and he drove the trepanners at Newstead. The trepanners had got valves on, and

*An underground roadway at Annesley colliery, 1973. (The National Archives, Crown Copyright)*

they were called poppets. They were to do with the hydraulics. I think they were like pressure release valves. When this trepanner broke down, it was a mechanical problem. Well, prior to 1974 and the power loading agreement, faces would sometimes run without fitters on. This chap's name was Keith Turton and he learnt the art, after watching fitters operate, to release these valves called poppets. He ended up not sending for the fitters, but doing the work himself, so he had the nickname of Poppet.

*A mechanic fitting cutter picks to trepanner at Gedling colliery. (The National Archives, Crown Copyright)*

Ken Lewis was a Welsh miner who came to work in the pits here in the East Midlands. Via his son, Goff, he provided me with a wonderful story about a miner they called Jack Eighteen-months. This was because he had lost half an ear in an accident, and *Half-an-Ear* turned into *Half-a-Year*, and thus Jack Eighteen-months!

Ken even put his memories of nicknames into verse;

A nickname would stick for the rest of your days
Clive Rose, they nicknamed him Petal.
Mr grandfather played a side drum in a band
So they gave him the nickname of Kettle.

Ron Rat caught rats as a hobby
His nickname was not very nice.
They thought up a name for Tom Thomas
And decided to call him Tom-Twice.

*Old headstocks at Kirkby colliery.*
*(The National Archives, Crown Copyright)*

Cabbage got his for tall tales that he told
He was a gardener devout
He went down his plot to cut cabbage one day
He shook one and two sheep fell out!

I remember our Overman, Jack Eighteen-months
How he got his name makes me laugh
A stone fell and cut his ear off, they say
So it left him an ear and a half.

A surprising number of miners put their thoughts into poetry. Some were commenting on their life, others the loss of a colliery headstock as a familiar landmark. All of them had a note of regret that the mines had gone.

# Chapter 4

# 'There Was This Character'

**Many of the miners I spoke to wanted** to tell me about characters they'd worked with down the pit. These were people who stood out in their memory as being eccentric or having a strong personality, or people who could be talked into doing something amusing.

Roy Mills started his career at Mansfield pit, but later worked for private contractors at Calverton and many other collieries. He told me, 'You do meet some characters underground. I had a deputy at Sherwood, and they used to call him the Colonel. He was a really big body builder and martial arts bloke. If he said it was Monday when it were really Tuesday, then it was Monday. He were all right with me because I used to work. He used to say, "There's workers and shirkers. The workers are all right but I'm here after the shirkers." We used to call him Colonel, because he used to march down the gate. He'd clout people underground. I've seen him clout a belt man, and the belt man says, "I'm going to see manager. The Manager phoned the Colonel up and said, "I've had a complaint, but that belt man needed a good kick up the arse anyway".'

*Men on colliery surface at Bilsthorpe colliery.*
*(The National Archives, Crown Copyright)*

Colin Wyld, who spent his entire career at Thoresby pit, recalled another miner with military aspirations. 'We had a guy called Joe Till who operated two conveyors from one place, a transfer point, in fact. Joe used to be in the army, but by now he had spent too long on his own underground. He used to pass his shift time marching with his hand-carved wooden rifle, which he carefully hid at the end of his shift. He used to bark his own orders and then carry them out. It was so funny to watch him if he didn't know you were there. He didn't ever see us, and it was rather sneaky of us to creep into his world. He lived it all as he had in the past.'

Roy Mills described another deputy he worked with. 'He was a German – Dieter Nasal. He was the face deputy, in charge of the face and the roadways. He was the hardest working person I've ever seen underground. He was fit. I give him his due, because deputies, you know, are not supposed to do no physical work, they're there for health and safety. But he did. He'd tell you what job wanted doing, and then he'd come assisting, doing the job with you. He was super fit, that's why we called him Robominer. If I carried two great concrete blocks on my shoulder, he'd got to carry three. Owt you carried, he'd carry one more. And that's how he was. He'd got to be better than you. Owt you can do, he'd got to do it better. A bit arrogant. Two concrete blocks are quite heavy to carry a good distance. A lot of folk could only carry one, because if you've only got a vest on, it'd rub your skin off. But he'd come down, with three on his shoulder. "Look at me, I can do three," he'd say.

'He used to come down the pit in just a pair of underpants with a great army belt on. Once in his past, he'd knelt over something, and a chain snapped and it hit him. It had split him open. He had this scar up his front, like a zip. A lot of people used to call him the Kraut, with him being a German, but the blokes who used to call him that were probably bone idle and they didn't like him because he were a hard worker. He was the hardest working man I ever seen. I could get stuck in, hard work never bothered me. My saying was: "Hard wock mecks good money". We'd go in weekends and make manholes on contract. Two of you could make one in four hours, and get two hundred and fifty quid. He used to go down half an hour before us, and drill the hole to blow it out. We say, "Leave it," and he'd say, "Give you a start." When we used to work with Deiter, how people used to talk to him was bad. I'll tell you something, I could imagine him being a handful if he ever turned on you, because he were all muscle. He was a good 6 ft chap. We used to call him Robominer, because he could lift owt.

'I've worked with some men from out Belper way. Ron Stanley, he was another character. I worked with him one Friday, and he had a tin sheet drop and it cut his hand. A great 'V' cut in his hand. Because he were going out that night, he just went home and stitched it himself with some catgut. They called him Doctor Ron after that. He were a right gangster, a great big chap. He'd been in trouble for dog fighting in his younger days.

'There was another lad at Ollerton, he used to work on the gate end, he was called Jimmy Stickbeetle. Not an ounce of fat on

Jimmy Stickbeetle, but he could wock. I bet he couldn't have been seven stone wet through. He looked like one of these insects. The muscle he'd got clung to his bone. Our deputy at Calverton, he was from Shirebrook, Mick Sherman. They used to call him Pressure Cooker, because his face was bright red all the while. He'd got high blood pressure, and he used to panic. We used to say, "Old Pressure Cooker's going any minute." He used to crack and go berserk. If there was a job where we'd been struggling, and we hadn't cut owt, he'd go, "What am I going to say to the boss in the morning?" You could tell when he was going to go on one, when nothing was going right. He was a good bloke though, Mick Sherman was.'

Mick Oscroft recalled a character from Shirebrook pit. He explained that as an apprentice, he alternated a three-month stint underground with a three-month period on the surface. 'It was

*Miners inspecting powered supports in Parkgate seam at Manton colliery. (The National Archives, Crown Copyright)*

*The pit bottom at Harworth colliery. (The National Archives, Crown Copyright)*

always good to have the three months in the fitting shops, especially in the summer for obvious reasons. The fitting shops were full of characters, never a dull moment. Me and a friend took a job into the welding corner to be welded up. The welder, who was ready for retirement, was a right laugh. His eyes were not so good and the steadiness of his hand was even worse. If it was a 'home job' you wanted welding, he would somehow know, and, as you walked in the welding corner, he would be rubbing his forefinger and the inside of his thumb on his overalls, to clean them. This was to make it quite clear to you or anyone else that he wanted a reward before he struck a blow. A cigarette would suffice. But the price would increase per the size of the job in hand.

'Anyway I laid the job across a six-inch vice on the workbench to steady it. I told the welder where it required welding and he commenced. A very bad habit of his was to lower the end of his rod to about five or six inches from the work piece, and just as you thought you had plenty of time to shield your eyes, he would stab his rod from that distance and arc up. He gave more welding flashes to more people than all the rest of Britain's welders put together. Anyway he began to weld my job. It was only a fairly short weld I required, so I was a bit anxious when it was taking a bit longer than I thought. Eventually he stopped welding. He lifted his mask, we unshielded our eyes and we couldn't believe it! He had welded the jaws of the vice together, all six inches of them. Did we laugh? Did we!'

Mrs F. Hallsworth put forward her own father, Arthur Ludlam, as a real character. Arthur was born in 1921 and worked at several local pits from the age of fourteen until he was sixty, mainly on the coalface, often working seven days a week. He is now 86 and still fit and alert. He had a reputation for fighting as a young man and was moved from pit to pit, but he always had a job because he also had a reputation as a good worker. 'When my sister, Yvonne, was about three years old, my father took her to Brookhill Pit in Pinxton, to collect his wages. The man paying the wages had a grudge against my father at the time, and told him to get to the back of the queue. This incensed my father, who went into the hut and the two men ended up fighting! Apparently, after the fight, my father went home leaving my sister behind, forgetting about her. Yvonne ended up being taken home on the bus by the son of Walter Scrimshaw, the local greengrocer. My mother was less worried about my dad leaving the child behind than him fighting again and risking the sack!'

Ron Booth recalled a number of characters from his time at Cinderhill. 'One chap, he was a part time chimneysweeper when he wasn't down the pit. He was called Sooty, but not to his face. One day, he went to the manager's house to sweep the chimney. The Coal Board had got some men doing something on the roof, and Sooty says to them, "When my brush comes through, give me a bang." But when his brush came through, they took it off. And when the next rod came up, they took that off. It kept going up,

*Coal pickers at Brinsley colliery, c.1912. (The National Archives, Crown Copyright)*

rod after rod, and he says, "Bleddy hell, what's going on?" When he found out, of course, his language was so bad that the manager's wife said, "Out! Don't come on these premises again." On another occasion, this chap called Len – I used to be an apprentice with him on the coalface – he wanted his chimney sweeping. He lived at Bulwell. So he asked Sooty, "Can you come and sweep our chimney Saturday morning?" "Course I can," he said. "What time do you want me round?" Len says, "Well, when I get home I always go up my allotment, but I'll tell the wife to let the fire go out so it'll be all nice and cool for you." They used to bank the fires up you know, they never let them out. He arrived on the Saturday morning with all his rods and blankets and everything was covered over. Up goes his brushes and he gets them stuck fast. He can't pull 'em down or push 'em up. So he scratches his head wondering what to do, and he decides to put

*Blackshale return heading at Babbington colliery, 1972. (The National Archives, Crown Copyright)*

some paper up the chimney and set light to it, in order to burn the bristles off the brush to bring it down. Of course there's gas in the bend, and ... boom ... off blew the chimney pot, and it's bang, bang, bang down the roof. All the soot came down into the house. Len's wife wasn't very pleased, so Sooty got sacked from there as well.'

Another unforgettable character of Ron's acquaintance was known as Bish. Ron told me, 'One trick I remember, when we were about fifteen, we were down the road, ganging, and there was a chap named Bish. He never bathed at the pit, and every other word he used was a swear word. He would not stop for overtime. Dead on time, say it was half past one, he'd be off home. So this particular day he was somewhere near the coalface and he couldn't leave the job on time so he was annoyed. I said to the other young lads, "Bish is down here." We could see his light coming along. He never wore a cap lamp, he always had a bucket lamp. He wore his helmet back to front. Puffing and blowing, he was. Well you know the air doors they'd got for ventilation, we fastened one up so he couldn't get through. Oh his language! About four of us sat there grinning, and he got a hammer and smashed it down. Broke it. So next morning, I came down the pit and Horace Berrisford the manager said, "Office!." He said, "Who nailed that door up?" Well, we didn't realise, being young lads, that those doors were for ventilation, gas coming out, escaping. We hadn't got a clue. He gave us a telling off, but I think he saw the funny side of what we'd done to Bish.

*A deputy at Calverton colliery adjusting his flame lamp for a gas test. (The National Archives, Crown Copyright)*

'And yet that man, Bish, came out the pit bottom one day, and I was just one of the young lads at Cinderhill. I been driving a scraper conveyor and sent down to the coalface. I hadn't got any water and it was boiling hot. Bish saw me and he said, "What's up with your so-and-so face?" I says, "I ain't got no water. I'm dying for a drink of water." He says, "Ain't you brought any?" I said "No", and so he gave me a drink of his water. He said, "Now remember this. You work in the pit bottom, don't yer? Never ever ever come down the pit without water because you never know when you'll want it." And from that day on, I always took a can of water down with me. I learnt my lesson. All that time, even when I was a deputy, I always carried water with me.'

Keith Staley's contribution concerns a man who became quite well known when the UDM breakaway union was set up. Keith said, 'Characters? Neil Greatorex was one. We worked at Bentinck together, I was on the committee and he was the president. He was a devil. He used to plague my eyes out. He used to get different men on the committee – especially one who fancied himself as a bit of a karate expert Anyway, he used to tell them tales about me that weren't true. But sometimes it went further than a joke. So, one day, I was that worked up, really steamed up, with all these men getting onto me about various tales he'd told them that weren't true – things that I'd done and said – deliberately to cause trouble, to wind them up. I got to the lamp cabin one day, and who should be just in front of me come up the pit but Neil Greatorex. And as soon as he saw me, he dropped his lamp, run through the baths, run through the clean part in his pit clothes, which he shouldn't have done, and locked himself in the Union office before I could get to him. I'm running after him, with a fourteen-inch long adjustable spanner in me hand, as I used to tighten the bolts on the ripping. I was shouting and bellowing after him, "I'm gonner kill you if I can get hold or you", cos at times he used to make my life a misery. In the office door was a little ten by ten inch window with reinforced glass with steel in and I was that mad I got the spanner and I smashed all the glass. He was cowering in the corner and I was shouting at him, "I'll kill you if I get hold of you, Gret." We used to call him Gret.

'Anyway, someone came up, I forget who it was, a deputy I think, and he says, "You're wanted immediately, the manager wants you." I had to go and see the manager after I'd had a bath, and Greatorex had to go in with me. The manager said, "What the hell was going off? I ought to sack you both straightaway." Eventually, after an hour's talk, he decided that it had been more or less a joke and that I really wasn't going to kill him at all. But I was mad at the time. He was a devil. Well, once, when he lived near the pit, somebody got him excited and he went down the garden with a double-barrelled shotgun and threatened a neighbour. The police come and arrested him. There's not many people know about that. I don't know how he got away with it. He was quite a hothead.'

## Chapter 5

# Pit Humour

**Although most pit humour was verbal,** sometimes it was in the form of practical jokes. David Lee told me about a wonderful trick played on a rather pompous workmate at Langwith colliery in the 1960s: 'The face I worked on at Langwith worked two shifts and we were on the day shift. The afternoon shift had a couple of real characters. One afternoon, they were coming to work from Mansfield Woodhouse, and they ran over a hare on the road. They knew that one of their mates kept ferrets, so they put the hare in a plastic bag and brought it to work to give him to feed to his ferrets. However, they missed their mate in the baths, and they didn't want to leave the hare in his locker because of the smell, you know, going off. This was mid-summer, and it was red hot in the baths with the sun blazing in through the roof. So they took the hare, still in the carrier bag, down the pit with them, because it was quite a cold pit, so there'd be no problem.

'They were walking up the gate on the way to the face, and the button man who ran the conveyors for them, a bloke who was getting on in years, was quite a big bombastic chap. They were always tormenting him, so they decided they were going to have a bit of a laugh. As they were walking up in front of him, one of them said, "Look there. There's a rabbit!" The other one said, "No it's not a rabbit, it's a hare." They had a bit of a scuffle, and they got the hare out of this carrier bag and they reckoned to catch it in a manhole. They pulled it out, jiggling it by its ears, jogging it up and down as if it was still alive. Then one reckoned to chop it, and said, "Well look at that, that's a good un". They were having a laugh and a joke. And Ted, the old button man, stood there absolutely spellbound. Of course, they all had a good laugh and a joke about it for a day or two after that. But the word soon got round among everybody else, and a couple of days later, I went in the local welfare for a drink, and Ted stood at the bar. I said, "I hear they caught a hare up the loader gate on Tuesday, Ted, didn't they?" Ted booms out, "Yes, that's right, they did. And if I hadn't have seen it with my own eyes, I'd have never have believed it".'

Bob Collier told me of a couple of silly examples of horseplay that went on at Newstead colliery: 'Newstead took people from everywhere, and some of my best friends came there from pits that shut in the '60s, like Bestwood. Newstead was a very humorous pit, and a lot of silly tricks went on. Working up the north side at Newstead, we'd got a car road and it was probably a mile and a half long. It was a three-rail system, and in the middle was a pass-by, where it went into four rails for the cars to pass each other. The interesting thing about the two-man riding cars that we went on was that one was called the 61 and one was called the 84. Those were the numbers of the two buses that came through Newstead village, and brought men to work. So they'd used the numbers of the buses to name the man riders. When the afternoon shift were going in and the day shift going out, because you'd take each other off on the district, it got to be a regular habit of throwing balloons filled with water and bags filled with stone dust as the cars were passing. Men going into work would be covered in water and in stone dust, as were the men going out, and they ended up looking like snowmen. Eventually it became that serious that it did get stopped.

'The number 61 car was open but the 84 one was covered. One trick was, as you were going uphill in the 84, people that sat at the front of the covered wagon would empty their water bottles out onto the roof. The water would run down the top of the roof and

*Mine cars in the underground roadway at Annesley colliery. (The National Archives, Crown Copyright)*

drop onto people lower down, in the man rider. I actually did this one day, and a very good electrician called George Geddis, a union committee man, was sat at the bottom end of the section – there were about four sections of two seats – and I was at the top end. I tipped my water bottle on the top, the water ran all the way down the top of the man riding car, and then straight down his neck. Of course, his outburst was unrepeatable. "Who the ..." Nobody said owt but he knew I'd done it. While the man rider was still moving, he climbed over all the four seats and the barriers, and he hit me the hardest punch that anybody's ever hit me with in my entire life. He says, "You're not going to do that again, Bob, are you?" I said, "George, you're gonna be my friend for the rest of my life." What I didn't know – because he had only just come to Newstead at that time – was that George was a semi-professional boxer. George would go to the Goose Fair in

Nottingham and take on the boxers in the booths, and beat them. He was a very, very good man and I ended up working with him. In the '74 strike, I went on picket duty with him. Me, Eric, and George Geddis were in the same team, coal-bagging in the '74 strike. What we did, we bagged the coal early in the morning, and other people took it out – we only did it for the pensioners.'

Keith 'Scouse' Pruden, described a humorous trick that was potentially dangerous: 'Tricks? Well, the old coal picks – the hand picks we used for picking out – they used to have little steel wedges to sprig them. Some of the older miners, especially during my training, used to take the steel wedges out and put just a little lump of coal or a piece of muck in the gap. Then they'd give you the pick and they'd say, "Scouse, start picking that, will you". And on the first hit, the head of your pick would come shooting over your shoulder. You'd be hitting it that hard, because as a young lad you'd be thinking I'll really give it one, you cracked the shaft into the coal, and the pick would come flying over your shoulder. You'd be stood with just the pick shaft in your hand.'

Scouse also told me of a funny accident that happened to a friend of his: 'We had these things called 'groshle' bars. They're the bars that we used to put at the coalface end. They're big steel bars, roughly about five foot tall. Basically they support the edge of the coalface until the machine has been in and turned round. They're over 140 lbs in weight, really heavy. We used to have to knock a steel pin through 'em, to link 'em together. I was working with a bloke called Taffy Lowtham on this particular occasion. I said, "Look, I'll take the weight, Taff, and you hit the pin out with the hammer." That's with the coal sledgehammer. I says, "You know what you're doing, don't you?" He says, "Yeah, yeah. I know what I'm doing." I said, "So I'll take the weight, and you belt the pin out." Well, these pins were on small lengths of chain, so the pins don't get lost. I lifts it up and Taffy smacks it one. But he smacked it so bloody hard with the hammer that it shot out like a missile. It swung round on its chain and hit him between the eyeballs. It knocked him spark out. He fell backwards into all the mud, still holding the hammer in his hands. I couldn't hold the bar

*The coalface at Sherwood colliery. (The National Archives, Crown Copyright)*

for laughing, and I ended up dropping it in the mud. Anyway, he were all right, but he had a bit of a cut, like, and a hell of a headache.'

Sometimes the practical jokes were carried onto the picket line. Geoff Peace told me of an hilarious incident during the 1984–5 miners' strike: 'We were on picket duty at Cottam power station. It was snowing like hell, and the lads were getting a bit bored, with lorries going by us, so we started to build a snowman. Now these two coppers come along in this car and what they did, they deliberately ran the car straight into the snowman and knocked it down. Now there's a certain irony in what I'm going to tell you, in as much as it never had publicity at the time. There was a stanchion nearby so the next snowman we built, we built it round the stanchion, hoping beyond hope that they would do it again. And the bloody idiots did, they drove into the snowman and smashed their car up.'

Bob Collier – and what a name for a miner – told me that at Christmas, there was always a tin in the canteen so that the men could put a donation in for the canteen workers. 'In the years between 1974 and 1984, they always used to put a tip box on the canteen table at Christmas. It was either kiss the canteen staff or you put money in the box. Now, Newstead canteen had got a young lady who we nicknamed Quasimodo and this particular day, she was on the canteen counter when the afternoon shift came in. It was either give Quasimodo a kiss or put money in the collection box for the canteen staff. That collection box got totally full up that day, because nobody would kiss Quasimodo.'

Geoff Peace told me a tale that went back to the old days: 'It was a three-shift cycle, you cut your coal, then you packed and drew off. Anyway this American mine owner had seen the coaling system, and he'd seen the cutting system, but he wanted to see the drawing off system. Now in them days, there was a chap called the Steel Bobby, who was a legacy of the old coal owners. His job was to spy on the men drawing the timber and the supports off, and make sure none of them were lost. Anyway they were going up the

*The canteen at Bentinck colliery. (The National Archives, Crown Copyright)*

gate, the manager's taking this American up to see this drawing off, and they see a stretcher coming down towards them with four people carrying it. The injured man had a crushed head due to a roof fall. The manager explained that the man would probably be off work for three or four months, and might then have mental problems. The American mine owner drawled, "If that happened in America, he'd be back at work in a month". "Oh," said the manager. Further up the gate, they saw two lights coming towards them. A man was helping an injured miner out of the pit. They told the manager that a chain had broke and the end had hit him in the testicles. The manager explained that if the man's testicles were badly damaged, they would have to be removed and the man would be off work for two or three months. Again the American commented that in the States, the man would be back at work in a month. "Oh," said the manager. When they arrived at where the Steel Bobby was trying to look busy, the manager pointed him out to the American and said, "See that man there. He has neither balls nor brains, and he'll be at work tomorrow!"'

Scouse Pruden recalled two events when the men had trouble with wildlife. 'When we put a new coalface in at Rufford, they had all the chocks built up on the pit top. And the chock legs used to have sleeves round 'em, to protect them from dust and stuff like that. We had an instance where you've never seen ten men move so fast in all your life. We'd installed this new coalface, put all the new chocks on, and we started coaling. It's getting warm, and we're all on the coalface – stuck in the middle of the coalface about 300 yards long – all on us hands and knees. Everyone's getting stuck into the job, the next minute there's lads going down the coalface like whippets. There's a wasps' nest stuck behind one of the chock legs, and the heat's kicked them up. When they're cold, wasps go dormant but when they're hot they don't half liven up. The wasps' nest was stuck behind the jacket. You've never seen ten men move so fast, all on their hands and knees, everyone climbing over each other. We absolutely flew off the coalface. Nobody'd go on the coalface for blinking weeks until the wasps had been got shut of.'

If dealing with lively English wasps wasn't bad enough, Scouse recalled an insect invasion that was far worse 'We had a load of chock nogs come down the pit one day – big four inch square pieces of wood. We used them for building packs with. These had come from Russia. Unbeknown to us, they have hornets in Russia. Snap time had come; we stopped the conveyor belt and got our snap. We're all sat round eating, and we could see something in the beam of us cap lamps, something just flickering around. We used to get moths and things like that. And my mate Colin, he had this bloody great hornet on his neck. He does no more than smacks it one, thinking it's a moth. It falls on the floor, we gets it like, and it was about two-and-a-half to three inches long. We captured it, and put it in an old tin snap box. A couple of days later, we were stacking these chock nogs up, and we were stood at the side of one, and suddenly we saw all the hornets were boring out of the wood. They'd come in the wood from Russia, and the heat from the pit is making them think that summer's here. They're flying all round the coalface. Everybody had a laugh, though. We had a lot of laughs up to the miners' strike, but things were never the same after that.'

Mrs F. Hallsworth recalled a poem that her dad, Arthur Ludlam used to recite to her when he worked at Bentinck pit. It went:

Hip, hip, hip, I'm off ter pit,
Wi' a cobble in me shoe.
Rip 'n tear, coss 'n swear,
Me 'oss it winna goo.
Gangers one and sixpence,
Corporals half a crown,
Deputies five and a tanner
For walking up an' down.
O'er boot tops in watter,
Up ter neck in sludge,
Nobday would wock at Bentinck pit
On'y such fools as us!

Keith Pruden was one of a number of miners who remembered unpleasant tricks being played, concerning the very necessary water bottle that miners always took down the pit. 'You never leave your water bottle on its own or it gets added to.' I think Keith was referring to urine on that occasion, but other things could be added too. Ron Booth told me, 'There was a deputy named Danko – I don't know why they called him Danko but that was his nickname. You'd see him come up the pit, he'd feel who'd got the most water in their pockets, and pinch some. A lot of deputies were like that, you know, they'd help themselves. So the lads thought they'd sort him. They put Epsom salts in, and that cured him.' The substances added could be much worse than either Epsom salts or even urine, though. Henry Richardson recalled, 'You used to get a lot of incidents of men stealing water. You took your water in and when you were down the pit it were precious. These blokes got suspicious that this youth were pinching their water. Anyway they got a Dudley, you know, a tin water bottle, and they filled it with gear oil. He come out of the gate, switched the light out in the dark, picked this Dudley up and he swigged it before he realised it were bloody gear oil. He supped half a Dudley, and it'd gone down his bloody gullet before he realised. I reckon it half killed him, he were never the same bloke after that, I can tell you.'

It wasn't just the water that could be interfered with. Scouse Pruden told me, 'At one time it was a no-no, to touch a man's snap. But I did know instances where you'd leave your snap, you'd go away and you'd come back to find it had been swapped. Some lads, if they didn't eat their snap, they'd chuck it on the stage loader or leave it near the button where the telephones were. When the bread went green, what people used to do was to eat your fresh sandwiches and put the mouldy bloody sandwiches in your snap bag. You'd go to open it and, oh God, it'd make you puke when you saw what were in it.'

Colin Wyld had a number of humorous memories from his 38 years at Thoresby. 'Three of us were placed at Thoresby colliery whilst awaiting permanent placement. We were put to work in the

*'Snap time', (drawing by miner Bill Bennett)*

yard, which was concrete covered. Our job was to load a cement mixer of awesome capacity. In fact we found that it required 200 shovels of sand, 200 shovels of gravel and four bags of cement for each mix. After two days of loading and watching the fat foreman driving the dumper truck, we realised who had the easiest job. We decided to fix him up, so we made a very sloppy mix of concrete for him to collect. When he offered his dump truck for a load, we gave him one which drenched him in concrete. We all wished that it would set and he would be left as a statue, but it wasn't to be.'

Getting one over the foreman was quite satisfying, but if a top ranking government minister was visiting, that was even better. Colin told me, 'When a high-powered visit from VIPs was on the cards, the mechanics, electricians and all who didn't have too much to do, were inveigled into making things in the mine look good. This is why when the Queen visits anywhere, she always says it smells of new paint. On this occasion, the cabinet minister Cecil Parkinson came to visit the newly developed Park Gate

*A visit to Bilsthorpe colliery by Sir Derek Ezra, NCB chairman. (The National Archives, Crown Copyright)*

seam. A brand new man-riding haulage had been installed, and the visitors were to ride in the open cars. The cars had all been cleaned and disinfected. Coat pegs were supplied for their heavy jackets. The jackets wouldn't be needed again until they came back, so they were taken to the top of the out drift. However, when the visitors came out, the jackets weren't there. They had been exchanged for old worn out ones, and the good ones had been spirited away. What an embarrassment for our leaders. No one ever breathed a word; even a lie detector test would not have found those jackets. Even better, on the back of the train of open haulage cars was a written message which read, "This is a better ride than Sarah Keays." '

Colin's other story concerns sending an apprentice on an errand. 'After promotion to shift charge engineer, I worked on the pit top. On the surface it was always the job of an apprentice to fetch any requirement from the canteen. I was a smoker at the time, and gave this lad, Jim Storer, some money and a request to get me twenty No 6 tipped. He asked, "What if they don't have any?" I told him to bring anything, meaning any brand of cigarette. Can you believe that he came back with a pasty!'

My final humorous incident was caused by a stranger to the area not understanding thc Nottinghamshire accent. The broad Nottinghamshire speaker will always render make as *meck*, take as *teck*, and water as *watter*. The short 'o' sound is very dominant too. Home becomes *om*, work is pronounced *wock*, thirty comes out as *thotty*. Bob Bradley recalled one occasion when a visiting geologist got thoroughly confused. 'We had an undermanager at Teversal called Ron Wood and he was a very broad speaking man. "Can you get 'old of the jollogist?" he said. "I want him to goo down there and 'ave a look. We've got a lot of dotty coal down there." I said, "Right, I'll sort him out." This new geologist came down. He wasn't a Nottinghamshire man, he came from the north-east. I said, "Right David, Mr Wood wants you to go on that face, because he says he's got a lot of dotty coal. Can you find out what the problem is?" Dave went off down the pit to investigate.

'Later in the afternoon, Ron Wood came into our office, and he and the deputy manager sat having a cup of tea. He says, "Where's that jollogist got to?" I said, "I don't know." They all cleared off and came back half an hour later accompanied by the manager, and said, "Hasn't he come out yet?" I said no. Four o'clock came and went, then half past four, and we were getting ready to go, when Dave comes staggering into the office. He says, "Well, I've been on every inch of that face, and for the life of me, I cannot find anything like dots in that coal!" Well, we all fell about in a heap, from the Manager down. He couldn't stop laughing. Dave didn't realise that dotty coal was our way of pronouncing dirty coal. He says to me, "After that, I'll come to you to get it translated before I ever put a foot on that chair again."'

# Chapter 6

# Stubborn Ponies and Hungry Mice

**A common sequence of jobs when a boy** went into the coal mines from school was for him to work on the surface until he was sixteen, often working at the screens where coal was sorted along with men who had been injured or were too old to work underground. Then they would work underground in the pit bottom, the area at the bottom of the shaft, helping the onsetter to push the full tubs onto the cage. Another job that came early on in a miner's career was to become a ganger on haulage; he became a pony lad. His job was to collect his pony from the underground stables, fit it up with collar and shafts, then haul materials to the coalface.

John Deaton reminisced, 'I worked at Wollaton pit as a ganger when I was sixteen. When I took my pony, Beven, out of the underground stables on a Monday morning, he would kick and

*Chocks on the coal face at Babbington colliery.*
*(The National Archives, Crown Copyright)*

rear up, flexing his muscles, after being in the stables all weekend. I had to take him to the bottom of the tailgate, twenty minutes from the stables, to get him ready for work by putting on his collar and shafts. The tailgate was the out-take gate to keep the air moving through the coal face from the main gate. One of his tricks was to run up the gate as I was trying to put his shafts on. He would stop after a bit and I had to try to creep up on him with my hand over my helmet light. Just as I got to him, he would be off again. He got wise and quite a few times went all the way up to the coalface, some 800 yards.

'My job with Beven was to supply the face with materials – steel roof supports, props and ring arches for the rippers on the night shift. The rails that the wagons ran on were a bit dodgy, and the wagons came off the rails quite often. I would have to unload some of the materials to lift the wagon back onto the rails, making sure I unhooked the pony first, or he would pull the wagon onto me. There was nobody there to help, I was on my own. The only person I saw was the deputy in charge of the district.'

Bob Collier told me a tale about a pony lad who had a special arrangement with his pony, an arrangement that went wrong on one occasion. 'There was a lad who worked at Newstead who'd got a reputation for being pretty bone idle. In the late 1960s he was ganging with the ponies. He'd got a lot of freckles so we called him Spot. He had this certain pony, I can't remember the pony's name now, but the pony was bone idle like him. If they did a gang inbye with material and unloaded, and they'd got a bit of time to wait before they went out, Spot would get down inside a gate and have a minute's nap. And the pony would get down at the side of him, and they'd both go to sleep. Now I'm not joking, this is the honest truth. The horse was renowned for waking up when it saw a light coming down the gate, and he'd always wake Spot, so he was never caught asleep. But this day, an under-manager came down the gate and found both Spot and his horse asleep. He kicked Spot, and said "What you doing?" Spot got up, kicked the horse, and said, "Why didn't you wake me up?" Nothing ever came of it because in the 1960s when this happened, you could turn a blind eye to it. This lad went on to work on the pipe staff, and in 1984 he was on strike all the while with us at Newstead. Unfortunately he died not long after the strike – he had a heart attack, and he was only in his early thirties.'

David Lee told me that sometimes the pony lads at Langwith pit used to have an illegal ride on the ponies. 'At the beginning of the shift the young pony drivers had to wait at the end of the loco road until the manrider had proceeded inbye, and then follow up with their ponies. Riding the ponies was strictly forbidden, but the loco road was over a mile long, so all the drivers used to ride illegally. I stood talking to my mate Johnny. Johnny had been to the local cinema the previous evening to see a John Wayne film, and was busy telling us all about it when the manrider moved off. The lads would normally climb on to some convenient tub or tram to mount the ponies, but this time, fired by his hero's exploits the previous evening, Johnny said, "I'll show you how John Wayne does it." He ran to the pony and attempted to vault onto its back, but he only half managed it. The startled pony promptly spun round and set off back towards the pit bottom stables, with Johnny desperately clinging on in Red Indian style with his arms round the pony's neck. They had only gone a few yards when they ran into the undermanager coming round the corner. There was a flurry of dust, and some harsh words spoken before Johnny resumed his long journey on foot.'

The ponies spent their whole life underground, coming to the surface once a year when it was the miners' annual holiday. They would spend a week – later, a fortnight – in a nearby field, enjoying the daylight, the sun and eating grass. Henry Richardson recalled, 'When it was the pit holidays, them ponies used to get in the fields prancing. And they'd take some bloody catching when it were time to go back. They knew they were going back down.' One amusing feature about this annual trip to the surface was the fact that when they went back underground, there was a period of several days when all the hoss muck was green. Underground the ponies ate chaff, but the grass in the fields had an effect on their digestive system.

Getting horses up and down in the cage could be a tricky job. John Deaton recalled, 'One job I remember was when I had to take a pony called Marshall – a big strong grey – into the pit bottom to ride up in the cage to go to a horse show in Wollaton Park. However, he never got there. As we stood there in the pit bottom, another pony was being brought down the shaft. His name was Sharper, and he was a real hothead. As he came down in the cage, Sharper began to kick out and neigh. He kicked the gate out, and fell out of the cage and down the shaft. The pony and the cage gates fell straight through the timbers in the pit bottom and into the sump. There was a tremendous crash, with dust and dirt everywhere. Marshall knew that something had gone wrong, and became very nervous. He never went up in the cage – I had to take him back to the stables.'

*Manriding cars at Selstone colliery. (The National Archives, Crown Copyright)*

Ernie Kirk had a story with a sad ending. 'Underground, at Gedling colliery, the man that used to collect the pony – it was his pony – he'd collect it from the stables in the morning, and he'd go inbye to where he was working. This particular day, on the top of the manriding road, there was an elderly fellow who'd worked with the pit ponies for the last nine or ten years. He was material handling, and when they'd taken the materials to a face, the pony would be resting. They would carry the feedbag – the nosebag – and they would put the feedbag over the pony's head. This particular day, he puts the bag on and left him. When he went back to tag the pony onto the tub, the pony bolted. It ran down this roadway, and there was air doors in the roadway, which you had to open before you went through. It ran straight at these doors – killed itself. And it was caused by a mouse. There was a mouse in the feedbag that had panicked it. Made it bolt and it killed itself.'

Ron Booth gave me a very interesting account of how he was trained to look after his ponies. 'The ponies I had were the biggest ponies in Cinderhill pit. I used to take one out one day and one another. The gate I worked in ran at the side of a seam of coal. That took most of the weight, so it wasn't too bad.

'I was told "You must lead your horse". Well, I thought nobody ever leads the horse. After walking a mile and a half you'd jump in the tub. I mean the pony would stop if there was anything in front of him, if there'd been a bit of a fall, or if he could see the rails not right, because they were very intelligent, these ponies were. Anyway he stopped, and he wouldn't move. I thought what's going on? It was the lowest part of the road, and I just couldn't get out. The roof was so low. I had to wait about half an hour before the electrician came up. He says, "They're looking for you." I says, "I can't get out of here." He told me there was some stuff across the rails, and that was why the pony had stopped. He got me out but I learnt a lesson there.

*Wollaton colliery, 63s district. (The National Archives, Crown Copyright)*

'And the next thing was, you lead your pony back to the stables, about two miles away. So I gets onto the main road and I could see the coal's coming out of the loaders and the empties are coming down. I'm in between them, and I turned round and said "Come on". The pony got in front of me, and he was gone. When I got to the stables, there he was waiting to go through the door, looking at me as if to say, "Where have you been?" So that was another lesson, hold onto him tight.

'There was one pony, my mate had him, and its name was Merry. When they went up to the coalface, they got to turn round to bring the empties back. But they didn't. They used to stop the conveyor belt, the pony used to step onto the conveyor belt, ride past the tubs, stop it again and get off again. The horses would actually ride on the conveyor belt!

'The pony stables were spotlessly clean. Absolutely spotlessly clean. This old chap, Ernie Drury, he was the ostler, and he loved them ponies. As soon as you walked in, he looked at them for any bumps and bangs. He'd say, "Come on, my baby," and comb them down like racehorses. They didn't have tails, of course, just a sheared-off little tail, and not a long mane, that was all cut back. He'd look in the hoofs for any loose plates. He really looked after them. I've been in the stables when somebody wants to take a pony out, and he'd say, "You can't take him out." "Why?" "Cos he was out last night." Some ostlers would say, "Take him," but Ernie would not let that pony go. He said, "He had a shift last night. He's stopping in. I'm in charge." And he really put his foot down. "You can take that one. He's spare." He was a good bloke.'

Not everyone treated the horses with such tenderness. Henry Richardson told me a story about how a horse – on the surface this time – was somewhat overworked by the younger miners. 'When I was fifteen I was on Creswell pit top, and they were doing a new mine car system, these massive big automated mine cars for carrying coal. It were our job to get them onto number two bank. We had to shove them up a big gantry, about one in three gradient. There were all young lads the foreman told to shove up the bank. It were like horse work shoving these up this gantry.

And there's a bloody shire hoss in the stable, featherbedded. This weren't a pit pony, it was a big shire horse that worked on the pit top. The ostler used to look after it like it was his own kid. So we went to the foreman and said, "We're slogging us guts out, and there's a bloody great hoss in there." Anyway he went to the ostler, and the ostler decided to let us have it. But he told us to move only one mine car at a time. Well we'd got one youth with us who'd worked on a farm, and he said the horse could pull three at a time. So we winched three of these mine cars onto this shire hoss. He was pulling away, and as he got near the top, all of a sudden he slipped. And it dragged him down, three mine cars and a shire horse on its side whinnying and neighing. It had slid about thirty yards down this gantry, this bloody hoss with its legs in the air. The ostler came running out. He took the hoss off us and he wouldn't let us have it no more.'

One of the ponies' tricks when they were underground was to steal the miners' food. They would walk past where the men's jackets were hanging and investigate the pockets for snap. If the snap wasn't in a tin or a plastic container, the ponies would help themselves. Apparently an onion would deter them, but they did love fruit. Keith Stanley laughed as he told me, 'The men used to hang their jackets up, with an orange or something like that in their pocket, and I've actually seen a pit pony suck an orange through the fabric of the coat. And when the miner came back for his orange, all that were left of it were a mosh. That pony had sucked all the juice out through the coat pocket.'

Ernie Kirk has already related the case of a pony which became panicked by a mouse in its nosebag. Mice were a constant problem down the pit. Originally they came down in the chaff for the ponies, where they soon bred and spread out from the stables to all parts of the pit. Bill Buck recalled that at Clifton, 'The mice used to eat the sandwiches. You'd be sat there, you'd turn round they'd be after your sandwiches. They come down in the tubs.'

Keith 'Scouse' Pruden remembered a method of catching the mice. 'We used to get mice down Rufford. We used to feed 'em. If you'd got a bottle, you stood it at a steep angle, chuck some bread

*Kevin Bell, Keith Pruden and Garth Hallan.*

in the bottom, then sit back. The mice used to get in but couldn't get back out again. We used to sit there and watch 'em. Their legs would be going like one-oh, and they wouldn't be able to get back out the bottle. After a bit, like, one of the lads would get a knife, cut the top off the bottle and let them go. We wouldn't kill 'em. Well, one lad did do, but I didn't like him doing that. I fell out with him over it. They come round when we were eating us snap. They never bothered me. A lot of us had sandwich tins, and they couldn't get in 'em so they weren't a problem. But if you had a bread bag snap wrapper, you'd shot it.'

Scouse may have been too good-natured to kill the mice but he was in a minority. Keith Stanley said, 'At the coalface at Newstead, we used to have a bucket dug halfway into the floor, full of water, with a plank on it with a crumb of bread on this

*Newstead colliery village. (The National Archives, Crown Copyright)*

plank. So, when the mice went over, obviously it tipped over like a seesaw, and it used to drown the mice in the water.' Eric Eaton added, 'We used to phone the score of how much coal we turned at the end of the shift, and they'd ask for the score, and then ask how many mice we'd caught. It was said they recorded that as well for the manager to see. But one of my mates – he were a comical lad – he phoned the score out this particular day. He said, "We've done 500 yards of coaling." They asked, "Have you got any mice?" "No," he said, "but we've fetched Keith Nuttall out the bucket three times." Implying he'd gone for the bread, like.'

Henry Richardson had an even crueller memory. 'I've known them shove mice in the shot hole on the coalface and blow the buggers up. There must have been a new act come out because they said they'd got to get rid of them all, they were causing disease and that. They came down Creswell, pit holiday fortnight, put stuff down, and I never saw a mouse from that day to this. We were infested with them but it cleared them out.'

Keith Stanley mentioned that at his pit, there was another underground creature – the stable cat. 'We had one at Newstead, in the stables underground. Because there were bags with food for the horses, they attracted the mice and so the cat had a constant supply of mice to eat.' Another miner mentioned seeing the pit cat riding on the conveyor belts, and commented what a strange sight it was to see a pair of eyes travelling along in the dark, rising and falling as the belt went over the rollers.

## Chapter 7

# Missing Fingers and Miners' Tattoos

**For workers down the pit injuries were always** a possibility. Almost all the miners I spoke with had some wound or other. Many had the blue marks sometimes known as miners' tattoos, where a minor cut, combined with coal dust had left a scar. Keith 'Scouse' Pruden pointed out some of his scars to me. 'I've got pit marks here on my wrists. I've ripped the fingernail off this finger, down the pit. That's one of the fingers I broke; it's never grown properly since. I've torn my bicep in two, so today I've only got half a bicep on my left arm. I've got a scar on my shoulder where a lump of stone fell out the roof, and hit me on the back. At the time, I thought it was just sweat that was rolling down my back, and it was only when a lad at the back of me caught me up and he says, "Scouse". He grabbed me by my shoulder, and I thought he were mucking about. I says, "Gerroff," and then I saw his hands were covered in my blood. It were bleeding profusely.'

Ron Booth told me, 'The blue miners' tattoos? I've got the odd one, scars here and there, bits and pieces. But, my dad, when he stripped off, all his back was blue. I thought to myself how did he do that? I didn't realise that some of these old miners when they cut themselves, they used to get the coal dust and rub it in to stop the bleeding and congeal the blood. Then when they got home they'd bathe it but they never got all the coal out, and the skin just grew over it.'

Scouse confirmed this old pit-lore first aid trick and added another one of his own. 'Another trick was, if you got indigestion, you'd get a lump of coal and suck it. That'd get rid of the indigestion.'

Roy Mills displayed his own pit wounds. 'I've had some stitches while I worked underground. I chopped my little finger off. I split my head and had stitches in that. I've had stitches in me knee when I cut that open. My mate dropped a piece of steel, and split all me knee. Had to have six stitches in there.' It wasn't the injury that concerned Roy, it was the fact that he'd got a driving test the next day, and he had to cancel it.

Keith 'Scouse' Pruden recalled, 'I was at Rufford for twelve, thirteen years doing development and coalface work. I saw a few sights. Lads getting buried, and I saw a haulage chain snap once – we used to have the old trepanners – and they use to have real thick links. The chain snapped, and this lad had just stuck his head over the side of the panzer to see where the machine was. As the chain snapped it cracked him one, broke his nose, blacked both his eyes, and knocked out three or four of his teeth. But you saw it, you dealt with it, and you got on with it. Some of the accidents I had at the pit included breaking my fingers, I've been knocked out, I've had lumps drop on me, been knocked out twice. I've had a haulage rope snap – kicked my legs from under me and knocked me out.'

This calm and philosophical way that miners dealt with their injuries was exemplified by a memory told to me by Bob Collier,

*Miners coming off shift at Rufford colliery in 1966. (The National Archives, Crown Copyright)*

an electrician at Blidworth pit. 'Just after I came out of my time in 1967, one of the first faces I went onto was 38s, up the north side. The men were on piecework, and us electricians were just on day wage, so we wouldn't knock ourselves out. We'd do our work in the gate but we'd try not to go onto the face unless we had to. I was always in the gate, and I became, if you like, the emergency ambulance man. I'd help people if they got injured. When people injured themselves, they'd come out and I'd put bandages on 'em. This day, a chap came out off the coalface to me. He said, "Bob, give us a hand here." I said, "What's up?" He said, "I've just chopped my finger off. I was setting a chock, I'd got a little bit of wood over it, a hole, and I left my finger there in my glove. It caught my glove, and it's just chopped my finger straight off." I said, "Oh come on then, sit down here and let me bandage you up." I watered it down, bandaged it up best as I could, and put his arm in a sling. "Right," I said, "do you want me to teck you out of the pit then?" "No," he says, "but will you go and fetch me glove?" The thing is, at that time, everybody had to buy their own equipment. They had to buy their own boots, belts, gloves, etc. And sooner than go straight down the medical centre, he asked me to go and fetch his glove back. I said, "Where is it?" He told me the number of the chock. I said, "Right. Do you want me to bring your finger back then?" "No," he said. "Sling it in the gob." And so I went up the face, lowered the chock off, got his glove out, shook his finger out of the glove, threw it in the gobbin at the back. I set the chock, went back out in the gate, gave him his glove, and I said to him, "Are you all right?" "Ah," he says, "I'm just feeling a bit wheezy, wobbly legs". I said, "Do you want me to come with you out of the pit?" – thinking I'd get out early – and he says, "No, I'll be all right now, I've been sitting down. Thanks for my glove." He came back to work later, minus the top

*Shearer on 93s face at Thoresby colliery in 1976.*
*(The National Archives, Crown Copyright)*

half of his finger. It just shows you. He thought more of his glove than resurrecting his finger.'

Talking to Keith Stanley and Eric Eaton in the Mansfield NUM office, they described how miners used to laugh at minor injuries. 'It were a regular occurrence for men to be missing fingers. You could go on every coalface and see men with fingers missing. But you know pit man's humour. If somebody had just blacked a fingernail or owt like that, the rest of the team would be in tucks, they'd be laughing their socks off. They'd be going about banjoing, and taking the mickey out of you.' Eric, remembering the pain of those black nail injuries, went on. 'It could be painful, that black fingernail, by God. I've seen blokes get an old razor blade, cut through the nail to let the blood out.' Keith continued, 'Yes, but they were laughable incidents, weren't they? A lot of miners had blue marks on their arms and face.'

Some injuries were a bit more serious than the black fingernail or superficial cuts. Keith continued, 'When anybody were seriously injured, all the lads would go in to help, and they did expose themselves to danger to get their mate out. I remember one, Terry Noone, Tet we called him, this slawm of coal had come over on him, a side of coal, and it had hit him at the bottom of his back. I shot over the chains to him and helped the coal off him, but he said, "I'm sore down here." So I lifted his shirt up, and he says, "Is it all right?" It had gone that deep that you could actually see inside him. I could see this purple thing moving in his back, some bloody internal organ. But I just shoved his shirt back and I says, "Neah, Tet, it's not that bad, it's just a scratch." I bandaged him up, and he went out of the pit. He was in hospital for two or three days for that.'

Roy Mills recalled a serious accident he witnessed. 'I've seen some bad accidents down the pit. My friend John – this was at Sherwood – he nearly died. He lowered a chock down and crushed all his head. Oh God almighty, we had to get this lump of stone off his head, a lump of coal as big as a car. Luckily, it'd had got wedged, so it'd hadn't smashed his head. We had to make a fulcrum with a steel bar to ease the lump up, and get him out. He were in intensive care for a good week. I went with him in the ambulance, in all my dirt. He caved all his skull in, and broke his jaw. He had to have steel plates to build his face back. He's still alive though, I see him out shopping now and then.'

Scouse Pruden emphasised how miners down the pit relied on each other, and how the lads rallied round to help a comrade in trouble. 'Everybody looked after everybody. We had instances like one lad got injured. He got his foot took off. It was his own fault, what he was doing was riding the panzer. His foot dropped into the inspection hole on the panzer and went straight through the inspection cover, whack, it took his foot off. We tried covering for him, but we couldn't. So he was off work, he was finished at the pit. We all had a whip-round. All the lads chucked money in. We had five-a-side football matches against other coalfaces, and other pits in the area who'd got to know about it. We raised money and it made enough to help pay his mortgage off. We made about twenty-five grand. They then found him a job in the control room.

'Another lad, Gordon Telfer, a young lad, he lost his arm, got it ripped off. He was working at Rufford colliery. He was cleaning out a conveyor belt end, and his shovel slipped. It went round the return roller and his arm went with the shovel. Now this lad was only nineteen year old, and he couldn't reach the stop cord for the conveyor belt. He was stuck. He spragged himself in such a way the rest of his body wouldn't go round the roller. He was there for about twenty minutes with his arm slowly being ripped off. Anyway, he gets on the tannoy to us. He says, "Can you send somebody out to the box end." Vic Clarke says, "Why? What's up?" He says, "I've had an accident, and there's something up with my arm." Not "I've had my arm ripped off." So Vic and a couple of other lads drove back out to the heading, to where Gordon was. He was sat on the conveyor belt and what he'd done, he'd rolled his overall sleeve up and stuffed it in the hole where his arm had been. He sat there, with his arm totally ripped off. I still see him round Mansfield.'

Ron Booth had reason to be eternally grateful for the quick response of a comrade when he was in trouble. Ron said, 'Once I was buried in the fast end at Bestwood. What happened, there was a coal cutter come round, and in the fast end it were real bad work. I was bringing the coal cutter into the fast end, and the roof was shaley and soft. It all collapsed and buried me. One of the chaps – and we was always arguing, but only in fun – and he had the sense to run back and scrape all the clay and clag off my face, else I would have suffocated. Anyway they got me out. I hadn't broken anything, just obviously shook. They said, "You're not walking out, you're going out on a stretcher." We gets to the pit bottom, and of course they were bringing me up in the cage. They said to this lad who'd got me out, "You can't come. You'll have to wait." He told me about it afterwards. He says to them, "Look here." And there in his arm was the biggest gouge. What had happened, when he went to clear my face, the stone had made this deep gouge in his arm. Eric Malpass, that was his name. Every time he used to see me after that, Eric would say, "How you're going on, lad?" and show me the big scar on his arm.'

Ernie Kirk had two bad accidents when he worked at Gedling colliery. He explained, 'I had both legs broken at separate times. The first one was when I was shot firing. I stemmed the shot hole, put the charge up and a detonator. I fastened the cable onto the end of the detonator, working at a height on the face of three foot or three foot six. The cable is coiled up, and you throw it. You're supposed to be twenty yard away when you fire a single shot, so you throw it, then you crawl to it, then perhaps throw it again to get twenty yards away. To be honest, you didn't always get twenty yard away – if you could get round a corner, you might only be eight or ten yards. I just turned to move away and this black pit wall, it just broke away. I jumped back, but I'd got a conveyor belt at the back of me. I jumped back as far as that but, as it come over, it got me just above the ankle. I'm talking about a piece of coal two yard long by three foot six, and two foot thick. I was absolutely pinned. I shouted out but, of course, the men had moved away because I was firing, so they had to come crawling back up and they eventually got it off me. I knew my leg had broken, I felt it go.

*Panzer conveyor (return end) at Selstone colliery. (The National Archives, Crown Copyright)*

'And then, the second time, I were very lucky. We were fully mechanised, no hand filling, everything was machine cut. I was on afternoons and I went up this tailgate leading to the face. The deputy on the district went up with me. All the men were already there on the machines, because they used to catch the man rider before we caught it. We got up to the lip – and you'd got to duck under the lip. It's a half moon of stone. They drill and fire and set rings as they advance. There were no belts, it was all panzer chains, and all hydraulics, one every yard. I shone my light down the face, the spotlight on my helmet, to look at the pan lines. The pan lines should be straight, but I could see that the chain was kinked. To line it up all you do is turn round to the chock and turn a handle and a ram comes out and pushes the pans into line. They hadn't been pushed over far enough in places, and I said to this deputy, Norman Wharton, "Have you seen this pan line? It

*'Setting the ring' (drawing by miner Bill Bennett)*

wants pushing over here." He was on his way down to the other end where the machine was, so he left me pushing over. I stopped near the tailgate end and I put a chock on push, but nothing happened. I put another one on push at the side of it, then a third one. They were only a yard apart so you can imagine the pressure. The piston of the ram comes out of a barrel, and it's fastened on the pans, so when the ram comes out the pans have got to go. I'm kneeling down, and the ram is underneath my foot. That ram should be covered by a thruster bar, for safety but, over time, they come adrift. The head of the bolt on the end of the ram got me at the side of my foot. The other side of my foot was stuck solid, so I was trapped and I couldn't do anything. It was still pushing, so I reached the chock to put it on pull back, but with having the next two still on push, it kept going. I squealed and Norman come rushing back. I was shouting, "Pull 'em back, pull 'em back!" He was knocking them off, some that he'd left on push as well. Eventually, when he got towards me with his last two, it released my foot. Oh dear, my foot was all crushed. I thought I'd lost it. If he hadn't have been there, I could have still been there now. I couldn't get to the tannoy to let anybody know because I was pinned. Absolutely pinned. If it had been a weekend, and I'd have been on my own, I'd have had it.'

As a surveyor, Bob Bradley saw lots of sights underground, and one that left an impression was seeing a fire in the gob. 'I had to escort the deputy director and production manager on the paddy wagon to this particular panel at Ollerton. It was some three and a half miles inbye from the pit bottom. We get on the face and there were these three places where the roof had dropped out, and they were trying to get supports in. But they couldn't get moving for all the dirt. It had buried the panzer, so it needed to be shovelled away by hand to get the machinery going, because there was too much weight on it. They were waiting for men to get going, but they couldn't because they'd got a fire, a gob fire. It was glowing red, and that's a sight that is really frightening. Men were shovelling in there for no more than five minutes at a time. Putting it into bags, smouldering coals and what have you. And they were

*Headgear at Ollerton colliery. (The National Archives, Crown Copyright)*

putting water to it, to try and put it out. But, of course, they couldn't spurt water into the gob because there's oxygen in water, isn't there, so in effect they would be feeding the fire. Water sometimes feeds a fire, as you know. If you don't put on a sufficient amount, it actually builds it. Strange that, but it happens. The heat from the fire was indescribable. We were sweating where we were standing, so you can imagine the heat for a man shovelling right on top of it. It were very frightening that, because had the air quantities changed, and they'd got a depreciation of oxygen and an increase in methane, there could have been an explosion. What you had to watch for is low pressure. That's why a manager always had a barometer, a pressure gauge, so that he could see whether you'd got an upper or a downer. If the air presure was constantly going down, gases from the waste would come out. If you'd got high pressure, it keeps them in. So, in places of low pressure, you'd always got a

*Head wheels at Ollerton colliery. (The National Archives, Crown Copyright)*

chance of methane gas coming out the waste, the gobbing. From that we learnt an awful lot. It took some stopping off, but they did it, and thank God for that. That was going to be a new development for me, in a different seam, three and a half mile from the pit bottom, a brand new district, which was going to go another three miles. I know they say "It's coal at any price," but that was one price I didn't want to pay. They stopped it off, and I thought "I hope they don't ever open that one up again".'

Colin Wyld recalled a fire at Thoresby pit, a serious fire, but fortunately no lives were lost. 'We had an underground fire on a panel called 69s on the Top Hard seam. It was the biggest mines rescue operation that has ever taken place there. The manager at that time was a gentleman called Bob Anderson, who looked after the sixteen or so rescue teams involved in restoring Thoresby colliery as a workable pit by sealing off the raging fire. It was a terrific undertaking. All the rescue teams needed feeding and the men ate so well that Mr Anderson used to say that the stoppings – the seals – had been erected "using coffee and sandwiches from the canteen". The cigarettes they smoked was another thing, but to get our pit back, the workforce would have paid themselves. Our own team was headed by an overman called Jack Hall ably assisted by Terry Clarke, dedicated men who risked their lives to reclaim a place of work for about 1500 men.'

Colin continued, 'A very serious incident with winding cages at Thoresby was when there were keps in use, steel blocks that were moved into a shaft under the chair bottom of the cage for it to rest

*Panzer conveyor (return end) at Selstone colliery. (The National Archives, Crown Copyright)*

on. It happened like this: signals were sent from the pit bottom and the pit bank to the engine man to commence the wind. However, the banksman, Des Knowles, forgot to withdraw the keps on which the chair was at, and slack winding rope was piling up on top of the chair. Now being a fit, strong, young chap, Des, in his panic, snatched the keps out via a long lever. The chair dropped like a stone, free fall for about forty yards. Then, suddenly, the slack in the winding rope ran out. The speed and weight of the falling chair caused the rope to pull out its capple, the device fitted at the end of the winding rope to couple it to the chains on top of the chair. The chair sped the 736 ft to the pit bottom, crashing through the heavy timbers of the pit bottom landing. No one was hurt, and, although it took ages to repair the damage, Des was not sacked. In fact, some years later, he became a winding engineman, a very good one too.'

## Chapter 8

# Death and Disaster

**Death was always close at hand down the pit.** Many miners had tales of having a friend killed, or being present at a fatal accident. David Lee was literally standing next to a man as he was killed at Langwith pit. Worse still, the man was a neighbour and a friend. David told me, 'When I first went on the face, the chap that lived next door to us – we lived at number 19 and he lived at number 20 – he was a very, very powerful man, both physically and mentally. I used to call him Uncle, but he were no relation to me. He were a face chargeman, and he were a top notch man. He was also a star workman, always did good jobs, and when I first went on the face, I went to work with him. I'd only been working with him four or five weeks when he got killed. He was stood at the side of me when he was hit by a piece of rock from a shot-firing accident. They were firing some shots in the heading and the shot firer went down one side slit and we went straight down. Uncle was stood at the side of me, and we were waiting for the shots. They fired and the dirt splattered round us a bit, but it didn't touch me. I said something like, "That's spread a bit," and I turned round, and he'd been hit. He wasn't dead then, he died in hospital that night. A piece of stone – it must have been the size of a marble – had gone straight through the top of his helmet and broke the top of his skull open. It unnerved me for quite a while. I even stopped off work about three weeks, while I decided whether to go back underground or not. But I did go back.'

Not letting a tragedy put you off going back to work was echoed by Keith 'Scouse' Pruden, who described how his sister's boyfriend lost his life. 'Bill McCulloch got killed in the Bilsthorpe disaster in 1993. He went out with my sister, and his mum was friends with my mum. When I went to the funeral, his mum turned round to my mum and said, "Get your lads out that pit because you don't want to go through what I'm going through." But everybody went down the pit and you dragged.' In other words the good pay down the pit meant that the young men would continue to work in mining, despite the ever-present possibility of disaster.

Ron Booth gave me an example of a fatal accident he had encountered at Cinderhill colliery. 'There was an old chap, who never ought to have been on the coalface but, if they were short, they used to put him on. There was a fall of coal, it just split before they cut it, and it buried him. He had knocked his head against the steel prop as he went down and that was what killed him. Once again, he was killed just like that and I'd been chatting to him and pulling his leg only ten minutes before.

'A lad I went to school with, his name were Tommy Snell, and once again we was cutting on nights and all the gate end come in. It just come in. They got them all out, as they thought, and then somebody says, "Where's Tommy?" "Oh, he went down there." Nobody could see Tommy and then the overman come round and said, "He's in there." We dug him out, but a girder had come just across his neck and killed him. He were just eighteen years old. That upset us a lot because we were at school with him. One of his mates never went down the pit again.'

Sometimes the accidents were caused by the folly of men

breaking the rules. Keith Staley commented, 'Miners had got a bad habit – cos we told them many, many times as it shouldn't be done – but they used to ride down the panzer which was now mechanised, so it was a steel chain. They rode crouching on their hands and knees. Although we pleaded with the men not to break the rules, they did. And as one chap was riding down, his knee pad strap caught under the steel flights, as we call them, and they couldn't stop the chain in time so he went under the machine. It made a right mess of him, killed him.'

Sometimes, the careless behaviour was committed by other people, not by the victim. Bob Bradley told me about an accident at Ollerton where he was the unit surveyor. He said, 'There was this face gently dipping inbye, about one in fifty, but it was quite steep outbye and then gently slackened off. Once again somebody was doing something wrong. It was usually somebody doing something wrong that caused 99% of accidents – it was very rarely you got a true accident in the pit. It was usually somebody taking a short cut or something. Invariably when it comes out, it's put down as an 'act of God' so that people get compensation. Anyway, this bloke Dougie Bowler was walking inbye. There were some lads up above and they were doing what they call hand tramming. They were supposed to have put lockers in the tub wheels and coupled the run of tubs to a direct haulage rope, but they hadn't. It was what we called a runner, tubs running away. Dougie was walking inbye and the tubs ran straight over him. He didn't die for three days. But it must have been awful. They ran straight over him. He'd no idea they were coming, because they were loaded. Down the pit, you know, you never ever hear a full tub coming. You hear an empty one because it rattles. A full one is solid, you see, so you don't hear it.'

Bob Bradley was also close by when a disaster at Sutton colliery occurred. 'This was in 1957 and just by chance we were doing some subsidence levelling on the surface nearby, on that particular morning. At twelve o'clock or thereabouts, the hooter at Sutton colliery continued to sound long blasts, whereas normally there was just two hoots to tell you it was knock off time. The long

*Pitheads at Sutton colliery. (The National Archives, Crown Copyright)*

blasts meant there was a serious incident, and jokingly we said, "Sutton's blown up." We were being facetious about it, but within minutes our boss comes hurtling up in his flying Standard Ten. We all piled in, and he said, "We've got to get back to the office and get ready because we've got to get back to Sutton colliery." We went back to Teversal, got changed, and then we went back up. One bloke had to stop in the office because he'd only got one pit boot, the other one was at the cobblers. He should have been with us but instead he sat by the telephone answering questions while we went. Perhaps he was lucky because we had to go, and we saw all the dead bodies and badly injured men coming out the pit while we were waiting to go down.

'We got down the pit about a quarter past one, and we went inbye to this particular district called Low Main 28s. There was a huge fall, a big hole as long as this room and about twenty feet high. That's an estimate. We couldn't measure it, because it were bitting as they call it, there were bits dropping off, so it wasn't safe to get underneath it. So we had to estimate it – as surveyors we were pretty good at estimating. What had happened was that they had driven what they call a duckbill wide heading. That was where they drove a gate, and extracted extra coal each side. They got the coal out, and they packed the dirt in the side, so it didn't need loading out. This was going to go a hundred yards, then open the face out, and carry on. Some bright spark decided – this was from Area HQ – "Ah, we'll leave about a hundred yards of coal, so we'll have that." So they decided then that they'd work it in another method. It meant that they'd got to take out the pack that was there, let the machine cut past them, and put the pack back on behind. So they were constantly taking it off, putting it on, taking it off, putting it on.

'Well, anybody in mining will realise that when you extract something, you've always got some fine cracks somewhere. Subsidence at the surface starts somewhere and that somewhere is behind the coalface. Anyway, the hole got bigger and bigger and bigger until they'd got this huge cavity. And the machine had gone

*Lump shearer at Pye Hill colliery. (The National Archives, Crown Copyright)*

down the face, but the panzer motor had got an electric cable going into it, a great thick cable. This great big piece of rock dropped out, hit this cable and chopped it off. The rock had brought gas down with it – there was 20% gas up there. Well at 20% of methane, it will not explode, it won't even burn, not enough oxygen but between 5% and 15%, you can have an explosion. And at 9.4%, it is at its most violent. So we think it had probably got 9.4%. There was a flash. Bang! It cleared the deck, killed five people, and injured numerous others. There were forty people in that district. There was a lad who was sat near the explosion, and he escaped. The undermanager sat nearby and he got all his face badly burned, and a chap who sat with him got killed. The people in the other gate were also killed, as were the men on the face and in the loader gate. It'd gathered force, you see. An explosion gets bigger and bigger and bigger. It thrives on the oxygen. That's why in the pit nowadays they have stone dust barriers. From an explosion, you get a bow wave; you get that before you get the fire. That would knock this barrier down and cause a stream of stone dust – pulverised limestone – and that would cut out any oxygen and thereby stop a fire. That's the idea. It did work in principle, but whether it works 100% in practice you never find out until it happens.

'Anyway it happened, and we were down there in two teams of three. The newly appointed surveyor had only been there about a fortnight, because the other chap had left. The natural progression in our group was from Teversal to Pleasley, or from Sutton to Pleasley, to a bigger pit you see. From Pleasley you went on to be a group man, like I did at Ollerton. This explosion happened, and we were down there until ten o'clock at night. The rescue team had left, so our only companions were two canaries in a cage. The part that I remember was the smell. It must have been burning flesh. The smell was slightly different to a rotten egg, slightly different to a dead rat, slightly different to a gob fire, but it was a smell of burning. I saw the print of a shirt emblazoned on the side of the roadway. Nearby, I found a piece of charred shirt, and the two buttons that fastened it. The rest of it had vaporised, leaving just a print of it on the side of the roadway. The other thing that fascinated me was when I picked up a 'bacca' tin. I opened it to look inside and there was silver foil, usually containing a twist of pigtail, but inside there was only a wisp of ash. The force hadn't melted the tin, but it had burnt what was inside it. That's incredible, isn't it? Those three things stick in my mind: the smell, the print of the shirt, and then this tobacco tin.'

In the Sutton colliery disaster, which occurred on 21st February 1957, five men died. They ranged in age from a sixteen year old supplies hand to a fifty-five year old ripper. A further fifteen men were seriously injured, including a seventeen year old belt driver and the deputy, who was fifty-two. The list of injured also included two brothers working on the coalface.

A disaster which involved even worse loss of life happened at Creswell Colliery in 1950. In the early hours of Tuesday 26th September, eighty men lost their lives in an underground fire. Henry Richardson, who started work at the pit a week after the disaster, told me that the day shift had noticed that the conveyor belt was badly scored, with a deep 300 yard groove in it, about six inches from the edge. It was still functioning, however, and a full shift of coal was carried without mishaps, leaving the repair for the night shift. When the night shift came on, the overman saw that a length of coal on one face had not been filled, and gave instructions that the belt should be run until the coal was shifted. By 11 pm, the groove had developed into a split, and just after 3 am there was a trailing end. At 3.45 am, the transfer hopper was full of torn belting and was on fire. At 4.20 am Mansfield Rescue station was informed, as was Chesterfield.

Ninety-nine men were trapped beyond the wall of flames and, of these, only nineteen escaped and only three bodies were recovered before the order had to be given to seal off the area. One man building the seals had to do so in the knowledge that the body of his father was one of those being sealed in. Twelve hours later, the seals were broken and another forty-four bodies were recovered. The fire was now found to be even more extensive than had been thought, and the area was resealed. It was almost a year before the remaining thirty-three bodies could be recovered. Many of the dead men were related, and the Dodds family from Creswell lost three brothers, John, Leslie and Ernest. A fourth brother, Joe, was on holiday at the time, or he too might have perished with his brothers.

After the final bodies had been recovered, Henry Richardson worked as a line lad in the area where the fire had been. Although the area had undergone several air changes and had been deep cleansed, the vile smell remained, indicating where the sealed-in bodies had lain. A memorial garden in Creswell village cemetery lists the names of the men who lost their lives in the pit disaster.

The price of coal could be at times a terrible one. We all owe a deep debt of gratitude to the men who mined it, at the risk of their health and their lives.

# Chapter 9

# Dogging On, Turning Coal and Surveying

**Many of the stories told to me by the** Nottinghamshire miners I met related to particular jobs they were doing down the pit, whether they were on haulage, shovelling coal, cutting on the coalface, working as a deputy or in the surveyors office.

Ron Booth told me, 'The first job when you went down the pit was in the pit bottom. In Nottingham, we called the job dogging on. When the empty tubs come down the shaft, you link about fifteen of them together, and then they went down to be filled with coal. After about five months, we were sent down the roads, down into the districts. I went with a ganger, an older miner, who'd come off the coalface. I had a week for him to show me the ropes, and then I was on my own. It involved picking your pony up, then walking it down about two and a half miles. When you got to your district – I was on 15s – you were hauling supplies. Later,

*Cage being lowered at Hucknall no. 2 colliery. (The National Archives, Crown Copyright)*

they put you on the coalface. Of course, when I worked there it was top belt loading, so you shovelled the coal onto the belt. I was a young lad shovelling away like bees wings. That was first thing in the morning when they were clearing where the cutter had gone by. Then, when it slackened off, you were waiting for the shot firer to start firing. There was what was called a fast end, and a bloke would throw his coal back to me, and I was throwing it on the belt. The first week, when I got home I came straight in – I can't remember having my dinner – and went to bed. My mates would come round at night, and I'd be in bed. I stopped there until next morning. And at sixteen years of age it about killed me, but I got stronger.'

The story of getting stronger, building muscle was echoed by Keith Pruden. 'When I was sixteen, I went underground on the haulage. I had to learn how to clip on, and how to slip a clip as well. I done that for nearly two years, and then, two weeks after my eighteenth birthday, I got sent to Mansfield colliery on 10 shillings, which was the training face. I remember we used to have to get a diesel locomotive that used to take us out to the face. God, it used to stink, as well. We got out onto the training face and it was all stable hole. With wooden props and boards. That was an eye-opener. We used to have to bore the ripping lip, and fire it. It was all on shovel, no machines, all on shovel. Hand filling. And God almighty, when I started off there, I had a twenty-eight inch waist, a thirty-two inch chest, and about six inch biceps. By the time I finished, I had a thirty-four inch waist, a forty inch chest and biceps of twelve or fourteen inches. I ended up with a hell of a set of arms on me. It built you up. You put a lot of meat on.'

Often an older miner would encourage a young lad to beef himself up. Keith continued, 'When I got back to my own pit at Rufford, I got put on development. I worked with a bloke called

*Headgear at Mansfield colliery. (The National Archives, Crown Copyright)*

*Coal cutters at Calverton colliery. (Ron Mills)*

Vic Clark, and he were fantastic, a real old collier, built like a bulldog. He were only about five foot six, but he were about five foot six across as well. He'd got massive arms on him. If you worked with Vic, you were on the A-team. I can remember when we were putting steel supports in, he says, "Scouse, can you lift a ring load?" He took me down the gate, and he gets hold of this ring load himself, stands it up, puts it on his shoulder, and starts walking with it. He said, "I'll take this one, you bring the other one." So I looked at him and I thought, no chance, I can't lift this. Any road, I went to lift it and I couldn't, I fell over. By this time, he'd come back, and I was on my hands and knees on the floor. He says, "Get that ring load on your shoulders." I said, "I can't lift the bloody thing." So bang, he hit me one. "Get it on your shoulder." "I can't." Bang, he gave me another ribber. He said, "I'm going to keep hitting you until you get it on your shoulder." I did get it on my shoulder but it nearly killed me.

'While we was having us snap, he says, "What you got there?" I says, "Spam." He says, "Let's have a look." I give him these spam sandwiches, and he looked at them, then chucked 'em over his shoulder onto the conveyor belt. He says, "You're not having that rammel. Here, have one of mine." Black pudding. Doorsteps. He says, "Don't bring any rammel like that if you're working with me, because you'll never be able to build up." Anyway, he brought me four every day and made me eat 'em. I worked with him for about two or three year, and I loved it. Brilliant. He was a great bloke.'

When Ollerton colliery closed, Roy Mills worked for twelve weeks for a private contractor, Glec Mining, helping to salvage equipment from the pit. 'We got some machinery out, the big electric transformers that were worth a bob or two. A lot of electrical stuff were expensive. We got it all out, nice and steady, nobody on your back. The pit had cooled right down when the production stopped. It was unbelievable how it had all cooled down.'

Then Roy was told that RJB – known locally as Budge after its owner, Richard Budge – had bought Calverton colliery and was planning to reopen it. He had a phone call from the manager, who knew Roy from his days at Sherwood pit. Roy decided to go there. 'For the first six week, we had to put all the belts back in. They were all stripped out when British Coal shut it. It had been stood a couple of year. They only set fifty men on for the first six to eight week. Then, when we got everything up and going, they set about a hundred men on. I was on face driving, the same again. I had six year there, and enjoyed every bit of it. Calverton wasn't very deep, it was a very shallow shaft. It was only sixty odd metres. When you go to Clipstone it was a thousand metre, that's why the headstocks there are so high. And Calverton were cool. You could wear a sweatshirt.

'We used to work at Calverton in the week and, at weekends, we used to go to other pits – Clipstone, Welbeck, or Thoresby – fetching stuff, because Budge owned all the lot. If there was an old face at Clipstone that had finished and it had some decent stuff

*Headstocks at Brinsley colliery.(The National Archives, Crown Copyright)*

that could be used, we used to go up Saturdays and Sundays to fetch it. We used to go down the pit, get it loaded up, come out and load it on a pit wagon, and we used to take it to Calverton. And on Saturday nights we used to go to Welbeck, on a big two thousand-metre drift there, with all this big heavy-duty structure. We used to go and get fifty metre of structure in twelve hours on Saturday night, and take it all back to Calverton. We even went down to the Channel Tunnel, because when it broke through, Budge bought all the air bagging there. He were crafty, Budge were, he'd buy owt cheap. I've been to Clipstone and fetched some big motors and gearboxes and Budge would send them away for a thirty grand refurbishment. To buy them new it would have cost £200,000, so he'd be saving £170,000. He did lots of that, you know. I was everywhere at weekends, especially Clipstone, which wasn't bad for me being just down the road. A lot of the lads I worked with at Calverton come from Nottingham way or Derby way.'

*Clipstone colliery headstock 2008. Despite their listed status, the Coal Authority has applied for permission to demolish them. (Steve Hextall)*

David Lee helped to build a drift into Shirebrook colliery. 'What happened at Shirebrook was rather interesting, a bit unique at that time. Shirebrook had had a massive production potential, but they was always struggling with shaft clearance. They couldn't get the coal out of the pit fast enough. So it was decided to drive a drift from the surface down into the workings, to bring the coal out by a belt straight to the surface.' David was actually classed as an assistant to the manager in charge of the drift, but because he'd already got his manager's ticket, he was actually working as the undermanager. Building the drift took two years, and then David spent another two years as an undermanager at Shirebrook. He informed me, 'Shirebrook and Pleasley, which is on the outskirts of Mansfield, were connected underground. Eventually,

I inherited Pleasley workings as part of my district at Shirebrook. So, periodically, I used to traipse through from Shirebrook pit bottom to Pleasley pit bottom which, because of the varying routes, was a five-mile trip. It used to be a full shift job to go and walk to Pleasley pit bottom and back again.'

Keith Staley moved from Derbyshire pits to Nottinghamshire, and he expected Nottinghamshire pits to be more modern, and with much higher seams. He was in for a surprise. 'In those days you followed the money round, whichever pit was making the most money. Eventually from Ripley, Denby Hall and Ormonde pits, I finished up in Nottinghamshire. In Derbyshire we'd gone from hand filling right through to mechanisation – but as soon as we got to Moorgreen, we had to go back to hand filling. The management there were in raptures because they'd got a fully trained crew of men that could start hand filling again. We went in the Low Main Tupton and it was a very low seam. In Nottinghamshire you were expecting seams of six and seven foot thick, but in the Low Main Tupton it was really very low. We weren't just crouched, we were lying on our side at times, to get our stint of coal out. Anyway, they eventually decided in their wisdom to mechanise it, and that was the worst decision they ever made. Because we got it all mechanised, and then some crackpot decided to get greedy and instead of moving one chock forward at a time, they said, "We'll move five chocks forward at a time," and it all caved in. They spent millions of pounds on buying chocks. We were putting huge wooden chocks in these holes, and we were hardly turning any coal at all.'

Bob Bradley was a mining surveyor. He was the unit surveyor at Ollerton, and eventually became the area surveyor based at Edwinstowe. However, in his early days he was a young assistant based at Teversall. On his first day, 1st April, he was introduced to everyone in the office. 'Three of them sat there with their backsides to this fire. One of them said to me, "What's that in the corner?" I said, "Well, it's a fire." He said, "That's right my lad. As long as that's up the chimney back when we come in the morning, you'll be all right." That meant I'd got to be there ten minutes before them to get the fire going. And, of course, at the end of the shift I'd got to damp the fire down, because it was an open fire, the floors were all bare wood, and all the plans were in pigeon holes within two or three yards of it. We used to have the newspaper delivered and the group surveyors were given the task of reading it – the public notices, etc – to see whether there were any large buildings going to be put up in a particular area, because it was up to him to sort out whether there would be any subsidence damage. One morning, I couldn't get the fire going so I put the paper up, like you did, to get it drawing. Of course, it whipped up the chimney and set on fire. So I was on a bollocking for two things: setting the chimney on fire and for losing the newspaper, which hadn't been read.'

Bob's time was not all office based; he frequently had to go down Teversal, Sutton, Silverhill and Pleasley pits. Bob

*Miners about to board the man rider at Silverhill colliery. (The National Archives, Crown Copyright)*

reminisced, 'My first trip down the mine was to help carry the legs and be the back-sight, as they used to call it. The back-sight was the chap who used to stand at the back with a lamp. At Teversal, we went to survey this 70s panel for the last time. It had become unworkable. The face was only about twenty yards long. The road round the other side was all winding, so there were quite a few sights to take. We came tromping back towards the pit bottom, and when we were about a quarter of a mile from the pit bottom, the boss said to me, "Where's that set of tripods? That set that were leaning against the coalface. You'd better go and fetch it." Well, I didn't even know where to go. He says, "Go down there, it's the third gate on the right. You'll remember you've been up that gate." So I went hurtling off, and there wasn't a soul about. As you can imagine, I'm looking at every shadow as I went. I'm running, frightened to death, found the set of tripods and hurtled back.'

# Chapter 10

# Strike and Division

**The miners' strike in 1984-85 caused a bigger split** in Nottinghamshire than any other coalfield. In Kent, Yorkshire, Scotland, North Derbyshire and many other coalfields, the miners wholeheartedly supported the national NUM call for a strike to stop the Thatcher government from closing down pits. They suspected – and later they were proved correct – that the plan was to close virtually all the country's pits, then put the few remaining ones into private ownership. It was also suspected that the Prime Minister was out to avenge the 1972 and 1974 miners' strikes which had indirectly brought down the Heath government. In some small coalfields, Leicestershire and South Derbyshire, the miners did not support the call for a strike. A few brave souls did strike but they were so few – just thirty in Leicestershire and even fewer in South Derbyshire– that the split was very one-sided.

In Nottinghamshire, however, the split was much bigger. In a vote held in March 1984, all Nottinghamshire pits voted to ignore the strike. Overall, the majority was 73% to 27%, but in many pits the vote was much closer. At Blidworth the pro strike vote was 39%, at Newstead and Sherwood 40%, at Bolsover 42% and at Creswell 45%. Bolsover and Cresswell are geographically in Derbyshire, but for historical reasons, are always classed as Nottinghamshire pits.

Despite the vote, 7,000 Nottinghamshire miners came out on strike, though this number fell to 4,000 during the year long strike. Seven thousand decided to support the national union, but 27,000 chose to work or, in the phrase used by the strikers, they chose to 'scab'.

It was very different in the 1972 and 1974 strikes. Bob Collier recalled, 'I was fully behind the 1974 and 1972 strikes. In fact, during the 1972 strike, my wife Julia, who worked in the offices at Newstead colliery, left work to have our first baby, so our first baby was a 1972 strike baby. Julia's dad, granddad and great granddad had all worked at Newstead colliery. There were four generations of them, and they had the right name for the job. They were all blacksmiths and their name was Smith.' Bob Collier could say the same about his own surname.

Bob continued, 'In the '74 strike, I was involved with picketing. We picketed at the power station at Wilford where Clifton colliery had been a few years previous. We had some very funny experiences there. In 1974, because of the electricity being off, old ladies would come up to the picket line and throw their tin of cold soup over us because they couldn't cook their lunch. On the other hand, we had other people bringing us food. The pork pie factory – Pork Farms – they'd got a factory round the back on the industrial estate, and they used to send us pork pies. We had a lot of support in '74.

'It was different in 1984 because we weren't united. We were totally united in '72 and '74, when it was about money. The '84 strike brought out the difference between the genuine man concerned about his job and his future and the greedy people who knew that what they were doing was wrong, but all they wanted was the money. And that is the sole difference between united strikes and I'm-all-right-Jacks. They thought that the government would never shut our pits. And what happened to them? They got stabbed in the back.

'During the 1984–85 miners' strike, I went on picket duty to South Normanton transport depot, where Winterbank colliery used to be. I was in the picket on the front line as the transport people were going by, and I saw a group of people throw a big building brick through the window of a police car. It happened about twenty yards from me.

'Later on, during the picket, there was a police charge. I ran away like everybody else. I was recovering from a broken leg that I'd done down the pit a couple of years earlier so I wasn't the quickest to run and I was caught by a policeman. He jumped on top of me, lifted my head up by my hair – I'd got hair at that time! He smashed my face down into the gravel, ripped open all one side of my face. I was took in the van, there was four of us: me, another branch official from Newstead, and there was a Yorkshire lad and a lad from Linby.

'We were taken to Alfreton police station and I was charged with damage to a police car. I believe the police thought it was me that had put this big stone through the police car window. It wasn't, I had not thrown it at all.

'I was charged with criminal damage to a police car in June, and I eventually came to trial in November, which meant I'd been kept off the picket line from June to November. It wasn't a bad thing in many ways because, along with Eric Eaton, we ran a flea market stall on Hucknall market, to raise money. I'd also collect food, and things like that. When the case came to court, they'd dropped the criminal damages charge to one of threatening behaviour. I was found guilty of that and bailed to keep the peace for one year. I was straight back on the picket line after that. Never did I throw a stone or do any damage to a police car, it was just that I was that slow running away I got caught. I was unfortunate.'

Keith Pruden suggested that one reason that the Nottinghamshire miners were reluctant to strike dated back to the closure of Teversall colliery in 1980. 'What happened when

Teversal colliery shut, we asked the NUM National Executive to help us to keep it open, and they said no. They wouldn't support us, and Teversal shut. That is where the rot for Nottinghamshire mineworkers set in. When they started shutting Yorkshire pits, Mr Scargill came to the Nottinghamshire miners and said, "You've got to support us." And we said, "Well, hold on, you didn't support us when Teversal was shutting so why should we support you?" That was where it stemmed from.

'But even based on that, I still walked out in 1984. Me, my dad and my brothers all went out on strike. It was hard during the strike. I eventually lost my house. I hadn't been married long, probably eight months, and it weren't a very good start to my married life. I was on strike for about four months, but then me and my two brothers all got letters from NCB, telling us that we'd broke our contract, because we were apprentices. So what we did, we went down to the strike committee with our letters. We went down to see Keith Scutt, one of the strike organisers, strike secretaries we called them. We showed Keith the letters, and Keith said "In theory, they can terminate your contract because you're an apprentice."

'So I went back, but we didn't like it. We did us normal working, and that was it. Then, after the strike, we had all finished. Gradually men started coming back to work. I was uncomfortable with it for a long time. There were lads who I used to talk to and have a laugh with, who had been on strike for the whole year. I wished I could have done the same but I couldn't because of the way the situation was. It made things awkward. It took five years in some cases before lads would start talking to each other. What they done, they even split a coalface team up. Half of the coalface team had been on strike, and half hadn't. What they did was create a coalface team that was made up of 100% strikers, who'd stuck it out. When they got back to work, it created a challenge between one coalface and another. There was even sabotage work. Conveyor belts were getting cut, even after the strike. They were deliberately getting slashed with Stanley knives. It killed off a lot of camaraderie.'

*Miners on the march. (Ron Mills)*

Geoff Peace told me, 'My biggest regret was that I was unable to convince the lads at Blidworth that they ought to get their bloody jackets off and get stuck in during the 1984–85 strike. Another regret I have, and I ain't told a lot of people this, at a mass meeting where there were about 500 people in the Welfare at Blidworth, one of the electrical charge engineers on the pit top asked me a question. "Geoff, can we win this argument?" I said, "I don't know, but I tell you this, you can't afford to lose it." My answer was based on the history of the steel industry, and what was happening all round us. The government was saving us while last, there was no ifs and buts about that. Thatcher wasn't going to act until she'd got the right team round her in the cabinet, until she'd got Nicholas Ridley working on his philosophy.

'There was another regret, I suppose. Some lads at Blidworth, I felt sorry for them. They'd been on strike, they'd give all they'd got, absolutely down and out, and they had to go back to work. I told our pickets, "Look, he's give all he's got, don't bloody shout and bellow at him." But unfortunately the lads wouldn't do that, and they gave him some stick, which hardened them against us as well.

'This particular family I've got in mind, there were two of them called Strang. Their father were called Willie Strang, and he were as blind as a bat. These two sons looked after Willie. This particular morning, about fifteen or sixteen bloody police vans came down Blidworth pit lane, and they completely surrounded the pickets. They were being a bit fiery, starting giving some lip. We didn't know they'd got two Black Marias round the back of the pit canteen. The next thing, they'd arrested twenty or so pickets and chucked them into these bloody Black Marias. Now, I went up to this inspector from Manchester who were running things. I put it to him, "You've lifted a lad there, granted he's had a little bit of lip, that's just the frustration of the moment, but his father is totally blind. He's totally dependant on these lads looking after him. If you take his son away, whose going to look after his father?" So he went and opened the back door of the Black Maria, and he said, "Strang, come on, go home and look after your family." Well he come out and he gave this inspector the biggest bollocking you've ever heard. "You four eyed ——" so they turned him straight round, back into the van. I thought, well, you can but try lad.'

Keith Stanley recalled, 'The scabs used to say that they weren't going to give in to intimidation and the bully-boy tactics from Yorkshire. Well, I can tell you from my point of view at Newstead colliery, I was going to work on the pit bus, on days, and it stopped at the gates. There was a peaceful picket from Yorkshire, stood at the gates. We went over and listened to what they'd to say. I went over into the canteen, and all of my face team came out that day and they stopped out. All this about we'd been bullied into it is a load of nonsense because we were asked quite clearly if we'd support. And we gave that support straight away.'

Interestingly, it was Keith Staley, a working miner and a UDM supporter, who mentioned the oft-quoted – and always officially denied – tale of soldiers being used to fight the pickets. He told

*The 'battle bus' or 'scab bus' used for taking working miners through the picket lines in the 1984/85 strike.*

me, 'In later years I found out as it wasn't the police that was taking charge of all the strikes, as we thought, it was the army. When I was in Derbyshire Royal Infirmary, the man in the next bed to me was a chief official of the ambulance brigade during the strike, a senior officer. One day he says, "Keith, I had to go to the Butterley Police Headquarters as part of my duties, and by mistake I went in the gymnasium. There was hundreds of soldiers changing into police uniforms. That's true, that is, it's very true." This ambulance officer, he had no need to lie. Years later, I told Neil Greatorex – he used to be president of our branch at Bentinck, before he got elected to the big UDM job – I told him about it and he said, "Oh, ah, we knew all about that." He added, "But nobody would admit it".'

Henry Richardson was the general secretary of the Nottinghamshire NUM – a full-time post – and he was a firm supporter of the strike, as was the president Ray Chadburn. He explained that as the strike went on, it wasn't other miners that were picketing the local pits, but the striking Nottinghamshire miners themselves. 'We had a lot of heavy picketing for about a fortnight, and there were a picket killed. Then when other areas started having trouble of their own, with men dribbling back, they had to keep their own pickets there. In the last six or seven months of the strike, it was our own men picketing the gates. You see, every time we went to court, our barristers would put our case to these judges who were about eighty-odd, all supporting the establishment. You'd see the judge nodding off, not bloody listening. Then, as soon as their barristers got up saying, "These Nottinghamshire men have gallantly marched to work through hordes of pickets," the judge would say, "Oh yes, we can't allow this sort of thing to happen in this great country of ours." It were all lies. For the vast majority, they were Nottinghamshire pickets and there was no violence. The lads weren't going to stone their own men, they just called them scabs and that was it. They'd got to work with them, got to live with them after the strike.'

Although Henry Richardson and Ray Chadburn were supporters of the strike, the other two Nottinghamshire full-time officials – Roy Lynk and David Prendergast – were opposed, and became leaders of the anti-strike element. Because of Henry's support for the strike, he was thrown out of his job, and became a leader of the Nottinghamshire striking miners. He recalled, 'What we did was form two strike centres, one in North Nottinghamshire at Ollerton, and one for South Nottinghamshire, at the AEU offices in Nottingham. Then we were able to send the money there. The strikers were getting no money from the Nottinghamshire NUM, and they were getting desperate. It had to come from the national NUM to me so that I could pass it on. Even then there was a cock-up. I got a message from national level that a cheque for £12,000 was coming to me. They couldn't send it to Berry Hill, obviously, because all finance had to go through Roy Lynk. After about a week I rang the national office up, saying, "I thought you were sending a cheque. We're desperate for cash." They said, "We've sent it." They said that they'd sent it to 120 Skinner Street, Cresswell. Well, I been left there two year, it was where I used to live before. They'd sent it to my old address. I had to get my brother to go down to that address and the envelope had been opened. There was a scab living there, and he'd opened the envelope. Of course it went all round the pits, "No wonder Henry Richardson's out on strike, he's being paid £12,000".'

The end of the strike didn't bring any relief for the strikers. The Coal Board would only recognise the new union, the UDM. Miners wishing to remain NUM were told that they couldn't have union subs stopped from their wages, like the UDM members could. The NUM loyalists had to visit men at their homes, persuade them to resign from the UDM and rejoin the NUM by signing a direct debit from their bank account. Henry Richardson recalled, 'We had about 6,500, to start off. But we couldn't get their subs. The board wouldn't let them pay their subs through their wages. The thing that saved us was direct debit. But you can imagine, we had to recruit again. I mean, we'd got us men but they weren't really our members until we got them to sign a direct debit. They weren't financial members. And miners are very reluctant to sign owt at all. You couldn't get them to join a bank at one time. But it saved us bacon, and we were able to keep a check on membership as well.' In some pits the NUM came close to getting a majority of miners, but, as Keith Stanley told me, 'The Coal Board were playing a game, a numbers game. Whenever we started getting close to a majority, they either closed the bloody pit, or they transferred some UDM members across.'

Despite all this and some vicious tactics at work – if three known NUM men sat together in the canteen, they were accused of "holding an illegal meeting" – the NUM managed to get a presence in all Nottinghamshire pits. There is still an active NUM office in a former shop in the centre of Mansfield and the Nottinghamshire NUM Ex and Retired Miners Association also works from there.

But the split remains. Miners who went on strke and miners

who worked (scabbed) still have nothing to do with each other. It was not unknown for the split to occur between members of the same family, brothers, or fathers and sons. Long-standing rows between neighbours can be traced back to their different attitudes during the strike. In July 2004 a 62-year-old man was shot dead with a crossbow in Annesley Woodhouse. Local people noted grimly that both the victim and the killer were former miners. The victim was a striking miner, the perpetrator a working miner. The division has never gone away. Like elephants, miners never forget.

## Chapter 11

# The Best and the Worst Of It

**When I asked the Nottinghamshire miners** what was the best thing about working down the pit, all of them came up with the same answer: the friendship that resulted from working with mates in often dangerous conditions, relying on each other for safety. Ron Booth said, 'You relied on each other, same as in the forces. If you're in a team, you've got to rely on one another. I've seen where people hated one another, probably it's known that a man was knocking about with his wife, and perhaps he wouldn't be able to give him a good hiding because he wouldn't be up to it. But any accident, he'd go and help him. Then, afterwards, they'd go back to arguing. But that was comradeship. Most of these men met up in the pubs at night or at weekends. They lived with one another, lived in the same area.'

Ron also spoke of how living in a mining community in Nottingham didn't lead to falling out with miners from other areas. 'I was born in Radford and grew up in Radford, and we were in a gang. Then there'd be a Hyson Green gang, and there'd be others, the St Annes Well Road gang, but we never went fighting and stuff. We'd go up the Nottingham Forest, where the Goose Fair is held now, to play football. We'd play against the Hyson Green lads, and say there was eight on us and they'd got ten, they'd say, "Oh, you can take one of ours." We never had this knifing or fighting. We used to go down the Meadows, to swim in the Trent with the Meadows lads. The Sneinton lads, we used to go roller-skating with them up the Victoria baths. We never had any problems with 'em whatsoever. The Meadows lads all worked at Clifton pit; St Anne's Well Road lads, they worked at Gedling; Radford and Hyson Green lads all worked at Radford, Wollaton and Bestwood pits. The Bulwell lads as well, we never had any problems with them.'

Ron grew up in a mining community within the city, and Keith Pruden lived in a mining village, but they both recalled the same feeling of community and comradeship. Keith said, 'I lived at Rainworth, and there was no such things as burglaries, car thefts, or damage. Everybody knew each other. I mean, I wasn't as pure as the driven snow but if I done something wrong, the village bobby would have a word with my dad. And then my dad would just crack me one. The police knew that if they told your

*Interior of Hucknall and Linby Miners Welfare, 1963. (The National Archives, Crown Copyright)*

*Back-to-backs at Newstead colliery village – note the washing lines. (The National Archives, Crown Copyright)*

dad or mam, it'd be sorted, with a better result that what the police could do.'

David Lee, who finished his mining career as an undermanager, told me about growing up in another village. 'My wife Irene and myself were both born in Langwith and have lived in Langwith all our life. We were both born on Portland Road, a row of terraced houses. I was born at number nineteen, in the middle of a row of thirty-two houses. Across from us there were another thirty, because they had a big shop on the corner so they were two less than us. And in between was the backs. On the backs there were a row of toilets and ash-holes as we used to call it where they kept the dustbins. I was born in 1939, so my memories are from the back end of the war. They'd only had electricity in the village – electric lights in the houses – from about 1936, and I think flush water closets had only come in at about the same time. So my parents, and anybody that were a few year older than me, can remember the night-soilers coming round with a horse and cart to empty the toilets. And I mean, we had no hot water. We used to have a stone sink in the kitchen about a foot deep with a cold water tap, and then the back fire, the black range with the oven underneath, and the boiler at the side that we used to fill with a ladle, and stone flagged red tiled floor. There was no television, as you know. You talk to people now about it and their jaw drops. The toilet, you used to set off two days before you wanted to go, because it was so far away. But the quality of life I think was better then than it is now. We never thought then that we were hard done to, anything like that. You used to have a bike and leave it outside all night, reared up outside the back door. It wouldn't last two minutes now, would it? And the colliery manager was God. I don't mean just at the pit, which it was anyway, I mean throughout the village. People would be reprimanded if they didn't dig their front gardens, and things like that.

'There were some good lads down the pit. The whole workforce, you know, they were much maligned politically and on the television, they didn't want to touch them with a bargepole, that sort of thing, but they're like the army. Nobody wants them until there's trouble and then it's like Rudyard Kipling said, "It's

*Thoresby Colliery Welfare Band in 1968.*
*(The National Archives, Crown Copyright)*

Tommy this and Tommy that and Tommy go away, But the saviour of his country when the band begins to play." In reflection, it was a hard life, I suppose, but we never thought about it as a hard life at the time.'

Roy Mills, a coalface machine driver, agreed that one of the best things in mining was the friendship, but he added a second benefit – the money. 'The best things about mining are the money and the crack with people. You couldn't laugh and joke like you did underground in a factory. They wouldn't stand for it, would they?' Roy freely admits that he was in mining to earn lots of money, and he was prepared to work hard and do lots of overtime to get it. 'The only way to meck your money when I was at Sherwood was on overtime. You'd got to stop two hours every shift or work Saturday and Sunday. On nights we'd stop Sunday night twelve hours, six o'clock Sunday night till six o'clock Monday morning. We could do twelve-hour shifts all week, lots of time, but we used to earn good money.

'After five years at Sherwood, I went to Ollerton, the best place I ever worked. I never earned so much money in my life. At Ollerton I met some very good workers. Ollerton had got good electricians, good fitters, and they kept the machines going. They used to call me the Golden Mole, because I used to live underground. In 1991, after my first year at Ollerton, I earned nearly fifty grand in a year. I was the highest paid miner there. I was driving on face, and I never stopped. I was up and down, just ploughing up and down all the time. It was like a river of coal going down the belt. I used to say to the others, "You'll have to eat your snap while we're wocking". The contracts they give you at Ollerton meant the more you cut the more you earned. The manager said, "Don't ever tell anyone how much you're getting. You're earning more than the Prime Minister." But he added, "I need the coal, and you're getting it me." I used to go to work Saturday night. Five of us, twelve hours Saturday night, we earned £500 each, taking seven or eight strips of coal off. You're talking a strip of coal being worth £18,000 after everything's paid for. We did seven or eight strips on a Saturday night, so the manager was

*Miners at Calverton pit. (Ron Mills)*

laughing his head off. He'd got five idiots to give up their Saturday night to cut him eight strips of coal, that's worth eighteen grand a strip. If you're prepared to throw yourself around in mining, you'll always get on. I used to have a manager phoning me up when I worked at Calverton, saying, "We've been stuck two days, we haven't cut enough to pay the wage bill. Will you come to work Saturday night for twelve, fourteen hours, and cut us some coal?" I used to go any time. It didn't matter to me. I earned more money. Mining's given me a good life. I always had nice holidays, I always had a new car. It's give me a good life. And now I've retired early, and I'm getting all my pension back that I've paid in.' Roy told me that his motto was 'Ard wock mecks good money'.

Asked what was the worst part of mining, Roy said, 'The conditions. I could stick heat but the thing I never liked was sludge. When you were paddling in sludge, after about an hour, your boots get covered in it, they get heavier, your legs ache. You know what it's like when your hands or your feet are in water for a long while, you go all crinkly. Oh I tell you, we used to put bread bags over our socks. Sludge is a horrible thing to work in, especially if you'd got to kneel in it or shovel it cos your machines are bogged down, I've spent many a Saturday night at Sherwood, the machine where we was heading out was on soft floor, all sludge. We used to have to dig all the sludge out. We used to get railway bolts, and make a platform. Dig out level, drop all these great railway sleepers – great big ones, take three of you to lift one – then you had to drive the great road header on to that. If you didn't, the more it cut in, the more it used to sink in the soft floor. You know when you're shovelling sludge, a lot of what you get on your shovel is watter. Obviously you've got pumps on the side, trying to pump it out so you can work. I tell you, it's down-hearting, depressing. All you can do is look at the clock and think I'm only here for twelve hours, half an hour travelling, you've only got eleven on it.

'I could stick heat, but a lot couldn't. These big guys, that's the last thing they want, humidity and heat. I was always lucky, because I was on the face, and the face drew the air through. You did get a little bit of breeze. But obviously you got the heat from the machine. That machine was switched on twenty-four seven. If you touched the steel plates on the machine, it'll bon your hand. You could fry an egg on it. I used to take two sausage rolls out the canteen and a tin of beans with me, and put them on for half an hour and they were red hot. Your transformers, when you start going onto big faces, the machinery's bigger. So you get transformers that are bigger. They could be as big as a room. And they throw a hell of lot of heat off. I've never worked in heat like that at Ollerton. You can't explain it.

'We had a lot of water at Sherwood, and some people were getting all scabs on them. A lad from Woodhouse, he had that psoriasis. What I used to do was to get some Brattis cloth, cut a hole in it and put it over my head like a poncho. We used to have

*Sherwood colliery.*

to do a lot of digging. When you're underground digging, you'd get a lot of water running on your back. It used to rot tin steel, so what can it do to you? It's all the acid in it. That's why it rusted the tins. This bloke I worked with, Pete, he'd got that psoriasis anyway, and when he got that water dripping on it, he had all bad sores. He had to go on a different job for a bit, out on it. Touch wood, it never bothered me, but as I say, I used to make a poncho out of Brattis cloth. So really, I wasn't getting a deal of water on me.'

Keith Pruden also mentioned the conditions at the pit where he worked. 'The coal face at Rufford, it was hard work, long hours, a very, very dusty pit. They got a lot of bad dust readings. You used to get layers of dust on your arms, your face, and your eyes would get caked in it. I was a lad of fifteen and a half when I went in the pit, I was underground at sixteen, sweating my nuts off, gobbling dust, working in mud up to my knees, carrying material through mud and water. And the heat, it was like 100 degrees, the temperature. And in the winter, you'd come from being in a red-hot environment into the pit bottom, and there you were freezing. It's where the air comes down the shaft. It was ever so funny. You know those penguins you see in Antarctica, it's just like that in the pit bottom in winter, because everyone used to huddle as close as they could for the warmth. You'd get behind the biggest bloke. All the big blokes would get shoved to the front because they're a lot broader than us, and everybody else would shield behind 'em.'

*Miners celebrating the millionth tons of coal in one year (1974–75) at Hucknall no. 2 colliery. (The National Archives, Crown Copyright)*

Keith also recalled the good times and the good pay. 'When I was seventeen and eighteen year old, I was dragging £25 a week take-home. I could give my mam a fiver for my board, and I could go out on a Friday night with a fiver in my pocket – and beer was 28p a pint. I could go down Mansfield, get absolutely ratted.' Keith also added a foolproof way of spotting a miner when you were out having a drink. 'You knew who the pit lads were because they all had black rings round their eyes because of bloody coal dust. It used to stick to your eyelids and your eyelashes. If you went in a pub and you saw a lad with black eyelashes, he was a coal miner. You stuck out like a sore thumb.'

# Index